Collins
INTERNATIONAL
PRIMARY



# Computing

## Teacher's Guide

3

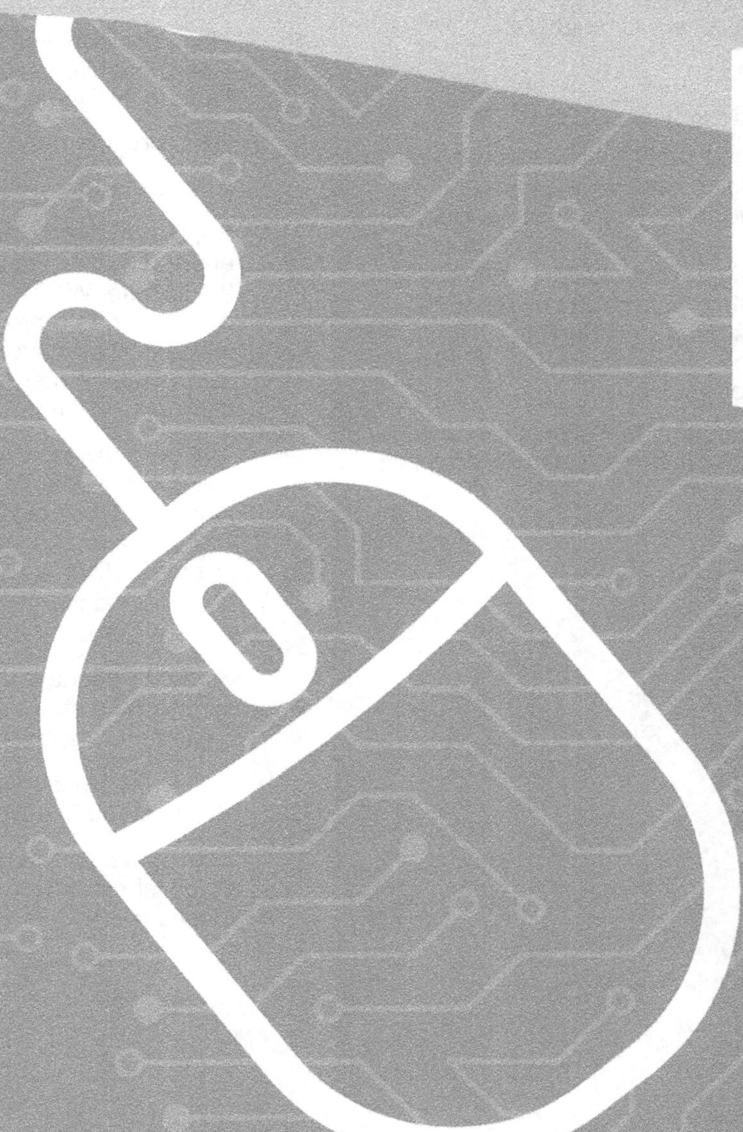

Rebecca Franks, Dr Tracy Gardner and Liz Smart

William Collins' dream of knowledge for all began with the publication of his first book in 1819.

A self-educated mill worker, he not only enriched millions of lives, but also founded a flourishing publishing house. Today, staying true to this spirit, Collins books are packed with inspiration, innovation and practical expertise. They place you at the centre of a world of possibility and give you exactly what you need to explore it.

Collins. Freedom to teach

Published by Collins

An imprint of HarperCollins*Publishers*
The News Building, 1 London Bridge Street, London, SE1 9GF, UK

HarperCollins*Publishers*
Macken House, 39/40 Mayor Street Upper, Dublin 1, D01 C9W8, Ireland

Browse the complete Collins catalogue at collins.co.uk



10 9 8 7 6 5 4 3 2 1

ISBN 978-0-00-868398-6



British Library Cataloguing-in-Publication Data
A catalogue record for this publication is available from the British Library.

Authors: Rebecca Franks, Dr Tracy Gardner and Liz Smart
Publisher: Catherine Martin
Product developer: Saaleh Patel
Project manager: Just Content Ltd
Development editor: Gemma Coleman
Copyeditor: Tanya Solomons
Proofreader: Tim Jackson
Cover designer: Gordon McGilp
Cover illustrator: Amparo Barrera, Kneath Associates
Internal designer: Sam Vail
Illustrations: Jouve Ltd
Typesetter: Sam Vail
Production controller: Lyndsey Rogers
Printed and bound by Ashford Colour Press

FSC www.fsc.org MIX Paper | Supporting responsible forestry FSC™ C007454

This book contains FSC™ certified paper and other controlled sources to ensure responsible forest management.

For more information visit: harpercollins.co.uk/green

**Acknowledgements**

Support materials and screenshots are licensed under the Creative Commons Attribution-ShareAlike 2.0 license. We are grateful to the following for permission to reproduce screenshots. In some instances, we have been unable to trace the owners of copyright material, and we would appreciate any information that would enable us to do so.

Python Software Foundation: Permission obtained to use screenshots demonstrating Python programming language features. PSF's License Agreement and PSF's notice of copyright, i.e., "Copyright © 2001 Python Software Foundation; All Rights Reserved" are retained in Python 2.0.1 alone or in any derivative version prepared by Collins (Licensee).

Scratch Foundation: Authorised usage of screenshots showcasing Scratch programming environment elements.

Scratch is developed by the Lifelong Kindergarten Group at the MIT Media Lab.

EduBlocks: Granted clearance for the inclusion of screenshots depicting EduBlocks interface and functionalities.

# Contents

# Teacher's Guide

## Introduction

Stage 3 Computing extends learners' knowledge and practical digital skills through the same six themes that are used in Stages 1 and 2. Learners encode and decode messages then move to developing a code of conduct for online communication. They develop practical skills through starting to touch type and developing a well-designed presentation. An underwater scene provides the context for learning about repetition and algorithms. Learners build an internet of things device that addresses a UN sustainable development goal. They create a micro:bit device that uses inputs and outputs. Learners document their school network using a visual approach of their choice. They develop their spreadsheet skills and apply them to a relevant project of interest.

### Overview

This book series uses six themes relating to computing and digital literacy. These are:

- Our digital world - Providing the tools and insights into safely navigating the digital world around us
- Content creation - Creating content effectively using a variety of software, from office tools to video production
- Create with code – Learning the fundamentals of programming and computational thinking skills
- How computers work - Discovering the specific technologies that make up a computer system
- Connect the world – Exploring how the world is connected through networks, the internet and the World Wide Web
- The power of data - Collecting, analysing and presenting data linked to real-world activities that create or empower change

Stage 3 has seven chapters that relate directly to these six themes, with two chapters dedicated to creating with code.

### Stage 3 Kit list

| | |
|---|---|
| **Chapter 1 – Our digital world** | • Coloured pencils, crayons or felt tip pens<br>• Optional: Large poster paper for sharing code of conduct<br>• Invite your school headteacher to watch the class deliver their code of conduct. |
| **Chapter 2 – Content creation** | • Computers with physical keyboards<br>• If you are using a non-QWERTY keyboard layout, then you should adapt the Scratch starter projects before sharing them.<br>• A prepared text file on your local network or cloud storage containing the example text<br>• Red and blue pens or pencils<br>• Projector connected to a computer<br>• Optional: a separate keyboard peripheral to use in a demonstration |
| **Chapter 3 – Create with code 1** | • Access to computers<br>• Scratch 3 – either the desktop app or online accounts |

| | |
|---|---|
| **Chapter 4 – How computers work** | • Scrap paper & pens<br>• Enough micro:bits for paired work<br>• micro:bits should be pre-programmed with the code provided<br>• Crafting materials<br>• Materials for a production line activity of your choice (e.g. building bricks, paper for origami)<br>• Identify some potential Internet of Things devices in the school<br>• Coloured pencils or pens |
| **Chapter 5 – Create with code 2** | • BBC micro:bits v2 (if using micro:bits v1 you could optionally add external speakers) or the micro:bit simulator<br>• Optional: a video of a dance routine |
| **Chapter 6 – Connect the world** | • Craft materials such as vegan marshmallows, dried spaghetti, string, glue, paper, painting materials, recycled plastic or boxes, and fabric cuttings<br>• Optional: a recent news report of a local network security issue<br>• Optional: photographs of network hardware in your school such as the server room |
| **Chapter 7 – The power of data** | • Computers with access to spreadsheets<br>• Optional: local residents for the showcase |

# Pedagogy

## Our overall approach

Our approach to teaching is based on the latest research in ensuring that computing classrooms are inclusive environments that foster greater participation of underestimated groups in computing and technology.

Each chapter is carefully organised to develop essential knowledge and skills whilst working towards the creation of a final project. The final projects are designed to boost creative skills and give learners the opportunity to make decisions and develop artefacts that matter to them. All chapters end with a showcase lesson that gives learners the opportunity to develop their presenting skills, whilst gaining valuable feedback on their work.

Every lesson plan states the key pedagogy that is used in that lesson. Stage 3 uses the following key pedagogies:

### Classroom talk

Classroom talk is the communication that happens in a classroom, between teacher and learner, learner and teacher, and between learners. Discussion and social interaction plays an important role in computing education. Throughout every lesson in Stage 2 you will find 'Discuss' activities. You are encouraged to incorporate topics that have been generated by learners and encourage them to share their ideas and perspectives and to ask any questions they have.

Benefits:

- Classroom talk provides an opportunity for multi-way knowledge sharing and building a shared understanding
- Valuing student voices and perspectives increases their sense of connection to computing

Read more about the research:
*Jenkins, 2017. Classroom Talk and Computational Thinking DOI:* (Website IntR1)

## Get creative

This key pedagogy is used to encourage learners to be creative with code or technology and provides them with the freedom to make their own design choices. Stage 3 provides a wealth of opportunities to get creative with technology, from designing their own avatars, to designing presentations, to building visualisations of their school network.

Benefits:

- Learners create projects that are culturally relevant to them, leading to greater engagement.
- Providing more choice in the project that learners create leads to a greater sense of agency, particularly amongst girls.
- Spending more time on the creative elements of computing leads to greater engagement of underestimated groups.

Read more about the research:
*Human et al, 2022. Gender balance in Computing - Evaluation of the Relevance Intervention.*
(Website IntR2)

## Real-world contexts

Lessons make regular links to everyday life scenarios where learners may encounter the concepts they are learning about. Links are also made to other curriculum topics. Stage 3 has a wide range of real-world contexts built into the learning, such as linking to their own use of online tools.

Benefits:

- Using real-world contexts aids learners' understanding by connecting computing concepts to their everyday lives.
- Real-world contexts demonstrate the value of computing, which engages a more diverse group of learners.
- Real-world contexts show the practical application of computing and make links to the world of work.

## Unplugged

Unplugged activities demonstrate computing concepts without using computers. Our unplugged activities are designed to relate directly to the technical concepts that learners are exploring. They follow a semantic wave (unplug, unpack, repack) while keeping closer to the technical concepts than other approaches to avoid confusion, particularly with respect to the international context. This means there is less need for 'unpack' and 'repack'. In Stage 3, learners understand algorithms by designing a dance.

Benefits:

- Unplugged activities have been demonstrated to positively affect learners computational thinking skills.
- Unplugged activities can increase learners' engagement with concepts and develop deeper understanding.
- Unplugged activities can be helpful when learners have limited access to computers.

Read more about the research:
*Chen et al, 2023. Fostering computational thinking through unplugged activities: A systematic literature review and meta-analysis DOI:* (Website IntR3)

## Develop practical skills

To be effective in computing, learners need to develop practical skills such using a keyboard and mouse and using IT tools. In Stage 3, learners develop practical skills in typing and using presentation software.

Benefits:

- Developing practical skills enables learners to become effective users of technology.
- Developing practical skills means that learners have more capacity to focus on the new computing concepts they are learning.

## Investigate

Investigate is the I in the PRIMM methodology for learning programming. PRIMM stands for Predict, Run, Investigate, Modify, Make. 'Investigate' involves learners interacting with code in a variety of ways to deepen their understanding. In Stage 3, learners investigate code that runs on the micro:bit to understand inputs and outputs.

Benefits:

- Builds a deeper understanding of coding concepts through interacting with code in different ways.
- Promotes discussion of questions that arise when interacting with code.

Read more about the research:
*Sue Sentance et al, 2019. Teachers' Experiences of using PRIMM to Teach Programming in School. DOI:* (Website IntR4)

## Working collaboratively

Working collaboratively means completing tasks in pairs or small groups. It includes group or pair programming. In Stage 3, learners experience the benefits of pair programming.

Benefits:

- Collaborative working can reduce cognitive load.
- Learners can complete work more quickly by working collaboratively.
- Collaborative working can lead to higher levels of engagement.
- Collaborative working reduces the number of physical resources, such as floor robots, needed for a class.

Read more about the research:
*Lewis, 2010. Is pair programming more effective than other forms of collaboration for young students? DOI:* (Website IntR5)

## Physical computing

Physical computing involves using tangible objects that children can interact with. In later stages, this can involve using motors and sensors. In Stage 3, learners design an internet of things device and write code for an interactive device that uses an input and output.

Benefits:

- Physical computing provides learners with additional context for understanding the results of their instructions or code.
- Physical computing is engaging and generates excitement and interest in the classroom.
- Physical computing provides a connection from computing to the physical devices that learners encounter in everyday life such as games consoles or supermarket checkout machines.

Read more about the research:
*Hodges et al, 2020. Physical Computing: A Key Element of Modern Computer Science Education Computer. DOI:* (Website IntR6)

## Role play

Role play activities involve learners acting out scenarios to reinforce concepts that they are learning. In Stage 3, learners role play creating objects on a production line.

Benefits:

- Role play activities allow learners to gain practical experience of an activity
- Role play is engaging for learners and helps them to remember important ideas

Roleplay is a well-established practice in teaching. It has not been extensively studied in computing education for children.

## Showcase

Showcasing is presenting a project to an audience. In Stage 3, every chapter ends with a showcase of the work that learners have completed.

There are lots of ways that you can carry out a showcase in your setting. You can choose an option most suited to your class. For example:

1. Each learner presents their project to the whole class.
2. Divide the class into small groups. Each learner presents their project to their group.
3. Split the class in half. Half of the class present whilst the other half act as the audience. Swap roles.

Benefits:

- Showcasing requires learners to reflect on their work so that they can explain it to others
- Showcasing develops important communication skills
- Showcasing enables learners in the audience to develop the ability to give feedback
- Showcasing provides an opportunity to celebrate the achievements of learners and the tangible outcomes of computing work

## Assessment

The books are designed to follow a mastery learning approach where all learners are supported and challenged to meet the age related expectations of the curriculum. Formative assessment should be used regularly to check the overall understanding of your class. Each chapter has a final project that is assessed summatively. Appropriate scaffolding is provided to allow for the majority of learners to meet expectations.

### Formative assessment in the student workbook

The student workbook provides opportunities for learners to apply their learning to various activities. The outcome of these tasks can be used as part of a formative assessment process to measure the learner progress throughout the chapter. Where a workbook question has a specific answer, answers have been provided. There are also 'model answers' to guide your assessment needs.

### Formative assessment and the lesson plan guidance

Each lesson plan has a guidance column. Where appropriate, additional support has been provided for the type of answers that you are looking for when questioning the learning. This will help to guide your formative assessment. If learners are struggling to understand the concept then you should revisit it with the support of the scaffolding and guidance provided.

### Summative assessment and the project rubrics

Every chapter has a final project that the learners are working towards. Each project has a project brief with requirements for learners to meet. The projects are then summatively assessed using a rubric. Each rubric has three columns 'Working towards', 'Meets expected', and 'Exceeding expected'. The majority of the class should be able to achieve the 'Meets expected' criteria. Some learners will exceed this expectation and go a step further than required by the project brief. A few learners may fall into the 'Working towards' category and should receive additional support and scaffolding.

Learners should have access to the rubric at the start of a project so that it is clear how they will be assessed. This also helps to support their metacognitive skills.

### Summative assessment and showcasing

Each chapter ends with a showcase which provides an opportunity for summative assessment. Teachers can complete the assessment rubric during showcase lessons. Teachers can ask additional questions of learners to establish their understanding of specific criteria.

## Hardware and software guidance

## Scratch 3

Scratch is a free graphical block-based programming language developed for young people and is the recommended tool in the Cambridge framework. It can be used to make animations, music, games, and more. The projects in Stage 1 use Scratch version 3.

Scratch 3 can be used online through a web browser on a laptop, desktop computer, or tablet. Alternatively, Scratch 3 can be used without internet connection by downloading and installing the offline version on a laptop or desktop computer. It is important to decide which option is best for you and practise using that method before a lesson with learners.

The online version of Scratch can be used with Scratch accounts. This enables learners to save their projects and open them in the future. Projects will save to their 'My Stuff' page. When using the online version of Scratch without accounts, learners will need to save their projects to the computer and load their projects from the computer.

Projects saved to Scratch accounts can remain 'private' to that learner, or be shared and viewed by others in the Scratch community. This option is chosen on a project- by-project basis.

With the online version of Scratch, you can set up a Teacher account. This is an alternative to a standard Scratch account and is used to create, manage and monitor learner accounts. A teacher account must be requested through the Educator section on the Scratch website. This should be done in advance of using Scratch in lessons for the first time as some preparation of student accounts will be required.

Some learner activities use a Scratch starter project. The starter projects are referenced in the 'You will need' in the Teacher's Guide. To use the starter projects online with accounts, you can give learners the project link and ask them to click 'remix' to save a copy to their account. If they are using the online version without accounts, they should use the link to open and then save a copy to their computer. If you are using Scratch offline, you will need to save the starter projects somewhere the learners can access and then ask them to save their copy to the computer.

Scratch can be accessed via (Website IntR7).

The following pages can be printed to give to learners. They detail the steps to open starter projects, save a copy, and reopen their copy in future.

## Scratch online – use a starter project with a Scratch account

### Start your project

- Open the link your teacher has given you
- Click 'See inside' 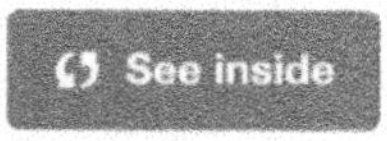

- Click 'File' 

- Click 'Save to your computer' 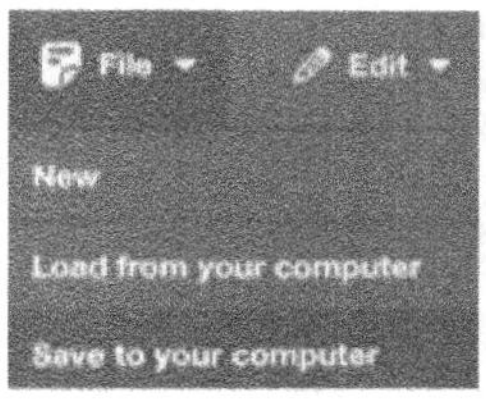

- Do the activity
- Save your project regularly

### Open your project in the future

- Go to the Scratch website
- Click 'Create' 

- Click 'File' 

- Click 'Load from your computer' 

- Find your file and click 'Open'
- Do the activity
- Save your project often

# Scratch offline

## Start your project

- Open the Scratch app 
- Click 'File' 

- Click 'Load from your computer' 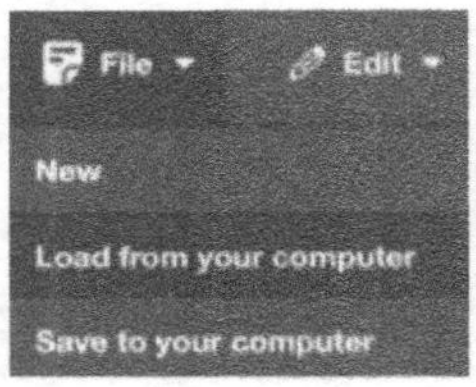

- Find the project your teacher has saved and click 'Open'
- Give your project a name 

- Click 'File' 

- Click 'Save to your computer' 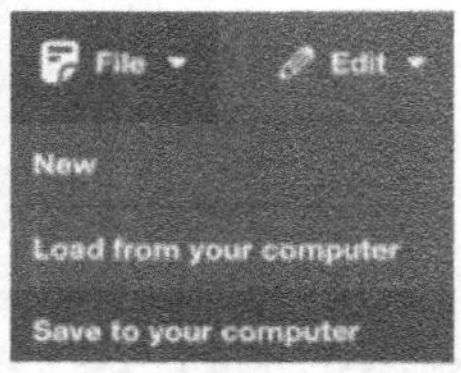

- Do the activity
- Save your project regularly

## Open your project in the future

- Open the Scratch app 
- Click 'File' 

- Click 'Load from your computer' 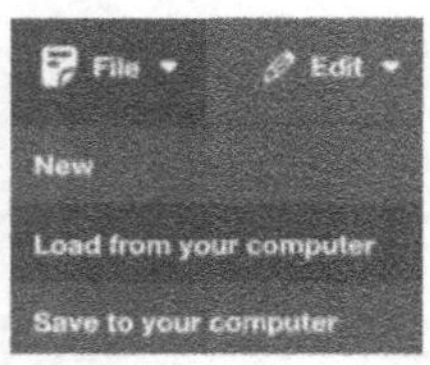

- Find your file and click 'Open'
- Do the activity
- Save your project regularly 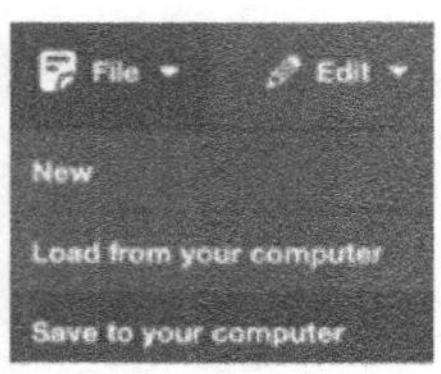

## Word processing software

Learners will use a word processor to practise their typing skills. You may already have word processing software installed on your school computers. The Teacher's Guide references Google Docs, a free web-based application.

Google Docs can be accessed via (Website IntR8).

Full support for using Google Docs can be found at (Website IntR9).

## Presentation software

Learners will use presentation software during Chapter 2, Content Creation.
You may already have presentation software installed on your school computers. The Teacher's Guide references Google Slides, a free web-based application.

Google slides can be accessed via (Website IntR10)

Full support for using Google Slides can be found at (Website IntR11).

## Spreadsheet software

Learners will use spreadsheet software during Chapter 7, The Power of Data.

You may already have spreadsheet software installed on your school computers. The Teacher's Guide references Google Sheets, a free web-based application.

Google sheets can be accessed via (Website IntR12)

Full support for using Google Sheets can be found at (Website IntR13).

## micro:bit

A micro:bit is a programmable device used in physical computing. It is the recommended device for Stage 3 onwards in the Cambridge framework.

Learners will use the MakeCode editor to create code for the micro:bits. MakeCode uses drag-and-drop blocks and builds on learners' experience with Scratch. Learners will test their code using the simulator in MakeCode before connecting their micro:bit to a computer and downloading the code to the micro:bit. Once the code is on the micro:bit, learners can use it via their computer or can power it from a portable battery pack.

There are two versions of the micro:bit: the newer version has a processor, microphone, 25 programmable LED lights, two buttons, a touch logo, GPIO pins, a speaker, a radio antenna, a compass, an accelerometer and a battery socket.

This book uses micro:bit v2, but options are given for micro:bit v1 or using the simulator in MakeCode.

A getting started video and text-based guide can be found via (Website IntR14).

There are also short CPD courses available for teachers via (Website IntR15)

## Online form software

Learners can optionally use an online form that contains questions they have developed. The Teacher's Guide references Google Forms as it is a free, web-based survey application that integrates nicely with the free web-based spreadsheet software Google Sheets. After creating a form, go to 'Responses' and click 'Link to Sheets' to set this up.

Google Forms can be accessed via (Website IntR16)

Alternative applications for online forms are available. If you use an online form application that doesn't link directly to spreadsheet software, you will need to manually input learner responses into a spreadsheet.

## Keyboards

The activities can all be completed using external, built-in or digital keyboards. There is guidance next to activities in the Teacher's Guide where a slightly different combination of keys might be needed depending on your setup.

One example of this is using Caps Lock on a physical keyboard versus double tapping the up arrow on a tablet. Or using 'Ctrl' on a Windows keyboard versus 'Command' on a Mac.

## Navigation

The activities can be completed using an external mouse, trackpad, or touchscreen. Learners will see the instructions 'Click or tap' in their Student's Books and have time to practise their navigation skills as they work through the activities.

 

# Chapter 1 – Our digital world

## Project: Create a code of conduct for communicating online and present it to your headteacher

## Chapter overview

In this chapter, learners will explore how messaging technology has changed over time and how to encode and decode messages. They will complete a variety of activities to discover the risks of revealing personal information online, and how to keep themselves safe and act appropriately. They will then be presented with a request from their headteacher to create a code of conduct for communicating online. Learners present this to their headteacher in the final lesson.

| Chapter 1: Lesson summary | | |
|---|---|---|
| **Lesson** | **Learning Objectives** | **Summary** |
| 1 | • Explain how messaging has changed over time with the introduction of technology<br>• Explain how a cipher can be used to keep information secret<br>• Encode and decode messages using a cipher | Learners discover how the use of messaging and communication has changed throughout history. They also discover that ciphers were invented to keep messages a secret. They then use a cipher to code and decode messages. |
| 2 | • Understand that people create online content for a purpose | Learners explore the different reasons why people create content online. They then choose a topic to create a short story about and present to an audience. |
| 3 | • Identify the risks of sharing personal information online<br>• Recognise how to engage safely in online spaces, including not sharing identifying information and media | Learners talk about the risks of sharing their personal identity online. They then design an avatar and a nickname that can be used to keep their identity secret. |
| 4 | • Model safe practice when engaging with online group messaging services<br>• Recognise that people can be upset by content written about them online<br>• Recall that content shared with other devices can be monitored | Learners explore the different levels of privacy their content can have when posted online. They also explore that messages can be hurtful and that they should always be kind online. |
| 5 | • Plan a code of conduct for online behaviour | The class collaborate on a class code of conduct. |
| 6 | • Showcase a code of conduct | The class showcase their code of conduct to their headteacher. |

<table>
<tr><th>End of chapter project</th><th>Create a code of conduct for communicating online and present it to your headteacher</th></tr>
<tr><td>Example ideas that learners could design:</td><td>
</td></tr>
</table>

# Chapter 1 – Our digital world

## Project: Design a robot to help someone who works at your school

| Name | |
|---|---|

| | Working towards (1) | Meets expected (2) | Exceeding expected (3) | Score |
|---|---|---|---|---|
| **Sketches** | • The learner has written 1 or 2 rules | • The learner has written three appropriate rules | • The learner has written three appropriate rules with excellent detail | |
| **Final design** | • The learner has draw 1 or 2 images to represent the rules | • The learner has drawn three images to represent the rules | • The learner has drawn three images to represent the rules with a high level of detail | |
| **Everyday task** | • The learner stated a rule with some support | • The learner clearly stated their rule | • The learner clearly stated their rule using strong presentation skills | |

| Teacher feedback | |
|---|---|
| **Learner response** | |

| Chapter 1: | Lesson 1 Messages through time |
|---|---|
| **Learning Objectives** | • Explain how messaging has changed over time with the introduction of technology<br>• Explain how a cipher can be used to keep information secret<br>• Encode and decode messages using a cipher |
| **New key terms** | **cipher** – method or algorithm used to encrypt plaintext messages by following a set of rules<br>**decode** – turning a secret message into a message that people can read |
| **Key pedagogy** | Real-world contexts |

| Activity | Instructions | Guidance |
|---|---|---|
| **1** | • Introduce the chapter project to learners (Page 1). They will be working towards making a code of conduct for how they behave when they are communicating in online spaces. | • Introducing the end of chapter project helps to focus learners on an end goal where they will apply their knowledge of communication.<br>• Make connections to any examples of codes of conduct that your learners may be familiar with. |
| **2** | • Read the information about how communication has changed throughout time (Page 2).<br>• Express how vital it was for humanity to invent ways to record their knowledge and communicate through time and distance. Use the images as examples:<br>  • It is thought that cave paintings were used for humans to share knowledge about how to find food and work together as a team when hunting. This knowledge led to higher chances of survival because communities could learn from each other and their elders.<br>  • As civilisation progressed, papyrus was invented as a way to record knowledge on paper. This knowledge could then be stored and transported across distances.<br>• Use Discuss 1 (Page 2). | • If you are studying this topic in history, this lesson provides a good link to that.<br>• Learners may have used technology to communicate with family members using text messaging or video calls.<br>• They may have communicated with friends online in games. Use this opportunity to remind learners about staying safe online. |

| Activity | Instructions | Guidance |
|---|---|---|
| **3** | • Ask learners to complete (Page 1) Task A, 'The correct order'.<br>  ◦ Learners add numbers in the right-hand column to predict the order when the methods of communication were first used. | • Answers are provided in the answer sheet. You can use these as part of a discussion with the class. |
| **4** | • Linking back to messages through time (Page 3), to explain that people needed to keep their messages a secret. Because of this, people created ciphers.<br>• Define ciphers, explaining that they are a way of coding messages so that only people who have the code can read the messages.<br>• Show learners the coded message in the book: 8 5 12 12 15.<br>• Use Discuss 2 (Page 3). | • Give learners some time to think about this before pointing out that the code actually says 'hello'.<br>• You could spend as much or as little time as you like here, depending on the lesson time. You could provide clues such as 'What is the order of the alphabet?' and 'Could the numbers represent a letter from the alphabet?' |
| **5** | • Ask learners to complete (Pages 2 and 3) Task B, 'Decode the messages', and Task C, 'Write a coded message'.<br>  ◦ Learners decode the messages and then write their own coded message. They will then swap with a partner and decode each other's messages. | • Advise learners to keep their own coded messages quite short.<br>• Remind learners that their partner will be decoding their message so it should be kind and polite. |
| **6** | • Reflect on today's lesson. Ask 'How easy would it be for someone to decode messages if they didn't have the cipher?' (Page 3). | • This coded message would be quite easy to guess because the letters directly relate to the number order of the alphabet.<br>• Another way to make it more secure would be to pair each letter with a different number from 1 to 26 so that you cannot easy tell which letter a number represents, for example A = 8 and B = 12. |

**Build on this:**

As a class, you could create a more complex cipher that is harder to crack using mixed up numbers. Learners can then create messages that others in your class can decode. You could put up examples in a classroom display.

| Chapter 1: | Lesson 2 Content creation |
|---|---|
| **Learning Objective** | • Understand that people create online content for a purpose |
| **New key terms** | **blog** – regular place online where a person writes information on a topic of interest<br>**podcast** – regular sound recording of a conversation among a small group of people on a topic of interest |
| **Key pedagogy** | Real-world contexts |

| Activity | Instructions | Guidance |
|---|---|---|
| 1 | • Remind learners that they already know that people online might not be who they say they are.<br>• Remind learners that people use the internet for lots of different purposes. | • Learners covered these concepts in Stage 1 and Stage 2. |
| 2 | • Explain different purposes for creating online content (📖 Page 4).<br>• Use Discuss 3 (📖 Page 4). | • You might be able to think of examples that are more relevant to your learners based on your cultural context.<br>• This is a good opportunity to discover how much time your learners spend online and the types of activity that they take part in. |
| 3 | • Continue to explain that there are different ways of sharing information online (📖 Page 5). It could be through videos, blogs or podcasts, for example.<br>• Use Discuss 4 (📖 Page 5).<br>• Remind learners that they should never post anything online without the permission of an adult. | • This links into the Workbook activity that learners are about to do. Spend some time now ensuring that the majority of learners have an idea that they would like to share.<br>• Providing further examples based on your knowledge of learners will help here. |
| 4 | • Ask learners to complete (📖✏ Page 4) Task A, 'Share your knowledge'.<br>  • Learners plan a short video clip of a piece of knowledge that they would like to share.<br>• Give them time to practise presenting their video talk. | • Draw on learners' talents and experience.<br>• Link to the purposes of communication (📖 Page 4). The purpose of learners' videos is to share knowledge. |
| 5 | • Complete a mini showcase with the class of the different topics they have planned to share. | • This could be completed in pairs, small groups or to the class depending on time. |

| | | |
|---|---|---|
| 6 | • Reflect on learners' showcases.<br>• Ask 'How did it feel to share your knowledge with an audience? What did you learn from listening to the talks?' (Page 5). | • Make the link back to the purpose of communication. Did learners succeed in sharing knowledge with others?<br>• Mention that the creators of videos, blogs and podcasts plan their content and think about their purpose, just like the learners have. |
| **Build on this:**<br>Learners could provide feedback on performances and make improvements to their video talks. | | |

| Chapter 1: | Lesson 3 Engage safely | | |
|---|---|---|---|
| **Learning Objectives** | • Identify the risks of sharing personal information online<br>• Recognise how to engage safely in online spaces, including not sharing identifying information and media | | |
| **New key terms** | **avatar** – picture, image or drawing used to represent your character online<br>**nickname** – name that you can give yourself online so that people can't guess your identity | | |
| **You will need** | • Coloured pencils, crayons or felt tip pens | **Key pedagogy** | Get creative |

| Activity | Instructions | Guidance |
|---|---|---|
| 1 | • Remind learners that people online might not be who they say they are. | • Learners covered this in Stage 1 and Stage 2. |
| 2 | • Introduce learners to the concept of keeping their personal identity private when they are online (Page 6).<br>• Use Discuss 5 (Page 6).<br>• Read the text to say why it is a risk if you reveal your personal identity online (Page 7). | • The five things are:<br>  • her first name<br>  • her school (the picture)<br>  • her date of birth<br>  • her phone number<br>  • the town where she lives.<br>• Make it clear that these aren't the only examples of personal information. Other examples include home photographs, address or your current location. |

| Activity | Instructions | Guidance |
|---|---|---|
| **3** | • Introduce the concept of an avatar to learners (Page 7). Use the examples in the book to let learners see what one might look like.<br>• Ask learners to complete (Page 5) Task A, 'Design an avatar'.<br>  • Learners design their own avatars. | • You might want to draw an avatar for yourself on the whiteboard as another example for your learners to see.<br>• Make connections to any apps or websites that you use in school that use avatars. |
| **4** | • Introduce learners to nicknames (Page 7).<br>• Ask learners to complete (Page 6) Task B, 'Nicknames'.<br>  • Learners explore nicknames that are safe to use. | • Discuss why the other nicknames in the Workbook activity are less safe (they contain personal information). |
| **5** | • Ask 'Why is it important to not share your personal identity in a public space online?' (Page 7). | • Remind learners that a stranger might use the information to find them in the real world, which can be dangerous.<br>• You can also discuss that someone might use the information to pretend to be them online. |

**Build on this:**

Learners could create digital versions of their avatars using the paint tool in Scratch or another digital art tool.

| Chapter 1: | Lesson 4 Think first | | |
|---|---|---|---|
| **Learning Objectives** | • Model safe practice when engaging with online group messaging services<br>• Recognise that people can be upset by content written about them online<br>• Recall that content shared with other devices can be monitored | | |
| **New key terms** | **public** – setting on an online account so that anybody can see it<br>**private** – setting on an online account so that only a specific group can see it<br>**code of conduct** – list of rules for being kind to one another and staying safe<br>**emoji** – image that represents an activity or emotion; emojis can be typed into a computer using an emoji keyboard | | |
| **You will need** | • Coloured pencils, crayons or felt tip pens | **Key pedagogy** | Real-world contexts |

| Activity | Instructions | Guidance |
|---|---|---|
| **1** | • Remind learners that they should never reveal personal information in a public space online. | • Recap the discussion on not sharing personal information from the end of the previous lesson. |

| Activity | Instructions | Guidance |
|---|---|---|
| **2** | • Introduce learners to the concept that messages online can have different levels of privacy (📖 Page 8). | • There are four examples given. You could expand on these examples in the following way:<br>• **Private messages:** Link this to sending text messages from a phone to a parent, friend or relative. A direct message is usually private, but it is possible for telephone companies and internet companies to have access to the messages if needed.<br>• **Messages on closed, private group chats:** It is possible to set up a private group to share messages between more than one person. Again, these are typically kept private but companies can access the messages if needed. It is also common in friendship groups for people to share conversations from a group chat with others, so the level of privacy depends on how much you think your friends will keep your conversations a secret.<br>• **Public messages on open forums:** There are spaces online where you can ask questions and have discussions on a particular topic. These are often public spaces and can be read by anyone.<br>• **Adding comments to public online videos or images:** Comments added to videos or images posted online are public.<br>• Mention that once you have sent a message online, it is not in your control anymore. Someone else could share it. So be careful what you share and who you share it with. |

| Activity | Instructions | Guidance |
|---|---|---|
| 3 | • Ask learners to complete (Page 7) Task A, 'Levels of privacy'.<br>  • Learners decide if the scenarios would be private or public. | |
| 4 | • Show learners the illustration (Page 9). It shows some hurtful emojis as comments on a Scratch project.<br>• Explain that our words can hurt others.<br>• Use Discuss 6 (Page 9).<br>• Point out that comments made online are monitored and websites often have rules that people posting need to follow. These rules are called codes of conduct. They usually involve being kind and staying safe. | • If there is time, you could briefly show learners the Scratch community guidelines (Website C1L4R1).<br>• You could ask learners if they have ever read the Scratch community guidelines or if they can predict what they say. |
| 5 | • Ask learners to complete (Pages 7 and 8) Task B, 'Design a kind emoji'.<br>  • Learners design an emoji that represents being kind. | • There are three examples of kind emojis in the Workbook. You could also draw your own on the whiteboard as extra inspiration for your learners. |
| 6 | • Reflect 'Share your emoji designs from the Workbook with a partner. Ask for kind feedback.' (Page 9). | • Model giving kind feedback, for example: 'I like the way your emoji has friendly eyes.' |

**Build on this:**

Learners could make improvements to their emojis based on the feedback received. If additional time is available, they could give kind feedback on the avatars they created in Task A, Chapter 1.3, of their Workbooks.

| Chapter 1: | Lesson 5 Design a code of conduct | | |
|---|---|---|---|
| **Learning Objective** | • Plan a code of conduct for online behaviour | | |
| **You will need** | • **Optional:** Large poster paper for sharing code of conduct<br>• **Optional:** Ask the headteacher to join you for part of the lesson to give the project brief. | **Key pedagogy** | Collaborate |

| Activity | Instructions | Guidance |
|---|---|---|
| 1 | • Remind learners that they can create content online for lots of different purposes. Recall that they must never share personal information online. Remind them that online posts are monitored and that they should always be kind. | • Make connections to online communities that learners have access, to such as Scratch. |
| 2 | • Go back to the project brief for this chapter, 'Create a code of conduct for communicating online and present it to your headteacher.'<br>• Show learners the examples (Page 10) and explain that they need to choose their own rules. Their code of conduct must be different from the examples. | • Show learners the Scratch community guidelines (Website C1L5R1) again to give them an idea of the kind of thing that they might want to say in their code of conduct.<br>• If the headteacher is available, you could ask them to join your classroom and request the code of conduct. |
| 3 | • Ask learners to complete (Page 9) Task A, 'Choose three rules'.<br>    • Learners list three rules that they would like to have on their code of conduct. | • Remind learners about being kind to one another.<br>• Also remind them about not revealing any of their identity in a public space. |
| 4 | • Bring the class together with their three ideas from the Workbook. They could share their ideas with a partner or their table to see if they have any rules in common.<br>• Ask small groups to share their rules. Type these into a word processor for the class to see on the projector screen.<br>• Discuss if they have missed any important rules from the list.<br>• Take a class vote on whether they all agree to the rules. | • It is important that they all understand and agree to the rules. This evokes a shared responsibility for how they will act together online. |
| 5 | • Keep the chosen rules on the projector screen and ask learners to pick three rules to create an image for. The image must represent the rule. It could be an emoji character if they like.<br>• Ask learners to complete (Page 10) Task B, 'Create three images'. | • Make sure learners are clear that they need to create images for the shared class rules. This may not match the rules they personally suggested. |

| Activity | Instructions | Guidance |
|---|---|---|
| 6 | • Ask 'Share your designs from the Workbook with a partner. Ask for feedback on how well the designs relate to the rules.' (📖 Page 10). | • Remind learners of the importance of giving kind feedback when they discuss the designs. |

**Build on this:**

Learners could update their designs to improve the link to the rules based on the feedback they received.

## Chapter 1: Lesson 6 Showcase your code of conduct

| | | | |
|---|---|---|---|
| **Learning Objective** | • Showcase a code of conduct | | |
| **You will need** | • Invitation to your school headteacher to watch the class deliver their code of conduct<br>• Your class code of conduct printed or displayed on a screen | **Key pedagogy** | Showcase |

| Activity | Instructions | Guidance |
|---|---|---|
| 1 | • Let learners know that they will be showcasing their code of conduct to the headteacher in this lesson.<br>• Help them practise how they will showcase the code of conduct to the headteacher (📖 Page 11). | • You may wish to prepare lines for each member of the class.<br>• Practise the whole showcase a few times to help your class develop their confidence. |
| 2 | • Showcase to the headteacher.<br>• Listen to the feedback that the headteacher gives. | • Display the code of conduct so that it can be seen by learners and the headteacher. |
| 3 | • Ask learners to complete (📖 Page 11) Task A, 'Reflection'. They can share their reflections with a partner.<br>• Celebrate what your learners have achieved (📖 Page 12).<br>• Ask 'How did it feel to present your code of conduct to your headteacher? How can you make your presentation skills even better next time?' (📖 Page 12). | • Remind learners that they should remember to follow the code of conduct whenever they are online.<br>• Emphasise that they have created a useful code of conduct.<br>• You could display a printed copy of the code of conduct in your classroom. |

| Activity | Instructions | Guidance |
| --- | --- | --- |
| **Build on this:** Discuss the feedback that the headteacher gave the class and decide whether you need to make any changes to the code of conduct. | | |

# Chapter 2 – Content creation

## Project: Create a presentation on a favourite topic

### Chapter overview

In this chapter, learners will practise using a physical keyboard to enter and then edit text in a word processor. They will look at similarities between word processors and presentation software before using presentation software to insert text and images. They will create and present a presentation based on a topic of their choosing.

| Chapter 2: Lesson summary | | |
|---|---|---|
| **Lesson** | **Learning Objectives** | **Summary** |
| 1 | • Improve typing skills | Learners identify the home keys on a computer keyboard and use the correct finger positions for the home row on the keyboard. They practise their skills using interactive activities. |
| 2 | • State how to use shortcuts for cut, copy, paste, save, undo and redo<br>• Identify and use spell-checking and proof-reading tools<br>• Identify common features of on-screen layouts and icons across a range of applications | Learners type text in a word processor and use the spell-check tool and keyboard shortcuts. They will open presentation software and use the text tool to add and edit text. Learners notice that the icons and shortcuts are very similar in the word processor and presentation software. |
| 3 | • Insert images into a document<br>• Use a search engine to perform keyword searches | Learners are introduced to the project brief and choose a topic for their presentation. They safely search the internet for content and images that they have permission to use. |
| 4 | • Navigate storage hierarchy for their device<br>• Create a presentation with images | Learners navigate the storage hierarchy of their device to add an image to their presentation. Learners plan a presentation that uses different slide layouts, including slides with titles and images. |
| 5 | • Create a presentation with images | Learners recall their knowledge of text formatting in word processor software and improve the visual appeal of their presentation. They format and lay out images to improve their presentation. Learners practise their presentations in pairs or small groups. |

| | | |
|---|---|---|
| **6** | • Showcase a presentation | Learners open their presentations in full screen presentation mode. They give their presentation to the whole class and provide feedback on others' presentations. Learners celebrate their achievement of creating and delivering a presentation on a topic. |

| End of chapter project | **Create a presentation on a favourite topic** |
|---|---|
| **Example ideas that learners could design:** |      |

# Chapter 2 – Create with code

## Project: Design and build an app that feeds a character their favourite foods

| | |
|---|---|
| **Name** | |

| | Working towards (1) | Meets expected (2) | Exceeding expected (3) | Score |
|---|---|---|---|---|
| **Sketches** | • Partial explanation of inserting images from storage | • Correctly explains how to insert images from storage on your school devices | • Provides a detailed and accurate explanation of how to insert images from storage on your school devices | |
| **Final design** | • Partial success in creating a presentation on a favourite topic | • Created a presentation with three slides on a favourite topic using many of the techniques introduced in the chapter (images, text, positioning, rotation, styles, colour) | • Created a presentation with more than three slides, effectively using a wide variety of techniques from the chapter | |
| **Everyday task** | • Shows their presentation with a partial explanation | • Delivers an effective presentation explaining their topic choices and discussing each slide | • Delivers an effective presentation and can answer questions using technical vocabulary including slide, slideshow and font. | |

| | |
|---|---|
| **Teacher feedback** | |
| **Learner response** | |

| Chapter 2: | Lesson 1 Improve your typing skills | | |
|---|---|---|---|
| **Learning Objective** | • Improve typing skills | | |
| **New key terms** | **touch typing** – pressing keys on your keyboard to write without looking at the keyboard so you can type more quickly<br>**home keys** – the keys on your keyboard that your fingers should return to in between typing letters<br>**home row** – the middle row of keys on a keyboard, which often starts on the left with ASDF | | |
| **You will need** | • Computers with physical keyboards<br>• **Optional:** A separate keyboard peripheral to use in a demonstration<br>• Scratch projects to be shared with learners during the lesson:<br>  • Home keys project (Website C2L1R1)<br>  • Home row project (Website C2L1R2)<br>  • If you are using a non-QWERTY keyboard layout, you will need to adapt the included Scratch projects before sharing them. | **Key pedagogy** | Develop practical skills |

| Activity | Instructions | Guidance |
|---|---|---|
| 1 | • Introduce learners to the new chapter and the project brief (Page 13).<br>• Recap previous knowledge of typing.<br>• Introduce the material on touch typing.<br>• Open a word processor and demonstrate how fast you can type.<br>• Show that if you type too fast, you start to make mistakes – speed and accuracy are important. | • In Stage 2, learners typed words into a document to label photographs of their school.<br>• It doesn't matter if you are not a fast typer, your typing speed and accuracy are likely to exceed those of your class!<br>• The key point is the trade-off between speed and accuracy. |

| Activity | Instructions | Guidance |
|---|---|---|
| **2** | • Introduce the material on 'home keys' (Page 14).<br>• Use Discuss 1 (Page 14). Show learners the bumps on a physical keyboard.<br>• Ask them to find the bumps or other indicators on their keyboard. Explain that these keys are identifiable by touch.<br>• Encourage them to look away and find the keys with their fingers.<br>• Ask learners to complete (Page 12) Task A, 'Improve your typing skills', where they will find the home keys on their keyboard. | • To find a comparison of different keyboard layouts, search 'visual comparison of different national layouts on a computer keyboard' on the internet. |
| **3** | • Demonstrate the Scratch Home keys project (Website C2L1R3). Type the letters as they appear, removing your hands from the keyboard between each letter.<br>• Share the link with learners and direct them to practise finding the home keys. | • Only lowercase letters are needed.<br>• Observe learners as they use the project and encourage them to:<br>  • look at the screen and not down at their hands<br>  • remove their hands completely from the keyboard between letters so they need to find the bumps again. |
| **4** | • Introduce the home row and correct finger positions (Page 15).<br>• Ask learners to complete (Page 12) Task B, 'Home row'.<br>  • Learners write the home row keys on a blank keyboard. | • Discuss the keyboard layout that you use in school, especially if it is not QWERTY. |
| **5** | • Demonstrate the Scratch Home row project (Website C2L1R4) and provide learners with the link.<br>• Ask learners to complete (Page 13) Task C, 'Typing speed and accuracy', and Task D, 'Home row words'.<br>  • Learners record their average speed at the end of each play of the Scratch project, and find as many words as they can type with only the home row keys. | • Stop the activity once all learners have had a chance to practise. They don't need to fill all the entries in the table in the Workbook.<br>• You can prompt learners to think of words for Task D by giving common word endings such as '-ad' for 'sad', 'had', 'dad'. Other words suggested might be 'as', 'has', 'ash', 'dash', 'lash', 'flash', 'flask', 'ask' or 'shall'. This activity could be done as a class discussion with learners writing the words in their books. |

| Activity | Instructions | Guidance |
|---|---|---|
| 6 | • Use Discuss 2 (Page 15). | • Hopefully most learners improved. Some may have got slower if they got tired or distracted.<br>• If you have time remaining, you can do the typing practice activity from the beginning of the next lesson. |

**Build on this:**
Find an online typing test and see how fast your students can type. Encourage them to return their fingers to the home keys.

| Chapter 2: | Lesson 2 Text and shortcuts | | |
|---|---|---|---|
| **Learning Objectives** | • State how to use shortcuts for cut, copy, paste, save, undo and redo<br>• Identify and use spell-checking and proof-reading tools<br>• Identify common features of on-screen layouts and icons across a range of applications | | |
| **New key terms** | **keyboard shortcut** – combination of keys you can use to perform an action instead of using a mouse<br>**presentation software** – program for creating presentations, with text and images, to be presented to an audience<br>**slide** – page in a presentation | | |
| **You will need** | • Completed Workbooks from the previous lesson<br>• The 'Example text' document for Chapter 2.2<br>• Red and blue pens or pencils | **Key pedagogy** | Develop practical skills |

| Activity | Instructions | Guidance |
|---|---|---|
| 1 | • Ask learners to open a word processor and start typing practice using the words they made in Task D (Page 13). | • Observe learners and remind them to return their fingers to the home positions on the keyboard.<br>• Remind them that this is something they should try to do whenever they are typing text on a computer, not just in this lesson. |

| | | |
|---|---|---|
| **2** | • Introduce the material on spell-check, grammar check and autocorrect (Page 16).<br>• Demonstrate how spelling and grammar errors look in the word processor you are using. Usually spelling errors are underlined with a red squiggle and grammar errors with a blue squiggle or two blue lines.<br>• Show how to accept a suggested correction.<br>• If your software autocorrects errors, show learners how this works so they recognise when their errors have been fixed.<br>• Ask students to type the sentence in Build 1 (Page 16), with the errors. Then ask learners to use tools in their word processor to fix the errors.<br>• Ask learners to complete (Page 14) Task A, 'Spell check and grammar check'.<br>    • Learners find spelling and grammar errors and mark them up in the style of a word processor. | • Tell learners that the word processor is not always correct. You could optionally show how to add new words to the dictionary if they **are** correct, such as a person's name. |
| **3** | • Cover the material on keyboard shortcuts (Pages 16 and 17).<br>• Demonstrate how to use the keyboard shortcuts. Make sure learners understand that they need to hold down the Ctrl key while they press the other key.<br>• Ask learners to complete (Pages 14 and 15) Task B, 'Keyboard shortcuts', and Task C, 'Fix the errors'.<br>    • Learners answer questions about keyboard shortcuts and cut and paste.<br>• Tell learners where to find the 'Example text' file on your cloud or local file system. Ask them to fix the errors in the document. Direct learners to save their work using the Ctrl-S shortcut and then close the word processor. | • Learners should have used these commands, using the menu or icons, already. They may need a reminder if they do not use them regularly.<br>• It can be quite tricky to master the timing of keeping the Ctrl key held down while tapping a letter key. Observe learners and offer assistance as needed. |

| | | |
|---|---|---|
| **4** | • Introduce the material on presentation software (Page 18) and tell learners which software they will be using.<br>• Ask learners to complete (Page 15) Task D, 'Our software", where they note down the name of the software they will be using.<br>• Do a quick walk-through of the presentation software and show that the menus, icons and shortcuts are similar to those in your word processor software. | • Remind learners that they will be using presentation software to complete their project. In future lessons, they will be taught how to add text and images to slides. |
| **5** | • Ask learners to complete (Page 15) Task E, 'Word processors and presentation software', where they answer questions about word processor and presentation software.<br>• Use Discuss 4 (Page 18) and discuss their answers to Task E (Page 15).<br>• Remind learners that learning new skills can slow you down at first but you get faster and learn how to do more things. | • Learners may also be aware of features that have not been covered yet such as images – images will be covered in the next lesson. |

**Build on this:**

Split learners into pairs and ask them to take turns in naming a feature that is the same in word processing software and presentation software. This could be shortcuts, icons or menus – how many can they name?

| **Chapter 2:** | **Lesson 3 Topic research** | | |
|---|---|---|---|
| **Learning Objectives** | • Insert images into a document<br>• Use a search engine to perform keyword searches | | |
| **New key terms** | **search engine** – tool that allows users to find content on the World Wide Web by typing keywords or phrases | | |
| **You will need** | • Access to a search engine with safe search features enabled<br>• Alternatively, access to a search engine on a specific website of your choice<br>• **Optional:** You could create your own example slide show to show learners. | **Key pedagogy** | Develop practical skills |

| Activity | Instructions | Guidance |
|---|---|---|
| 1 | • Re-introduce the project brief (Page 19). Show the example slides and a complete presentation project, if you have made one.<br>• Tell learners what kind of topic they are allowed to choose.<br>• Discuss what makes a good topic and ask learners to complete (Page 16) Task A, 'My topic', where they choose their topic. | • You could link topics to a subject being covered in another lesson or related to your local community.<br>• You might want to highlight the kinds of topic that are not suitable, if you think learners might choose a problematic topic (e.g. linked to a video game they are not old enough to play).<br>• Here are some ideas for what makes a good topic:<br>  • Choose a topic you know about or are interested in.<br>  • Choose a topic that has interesting pictures.<br>  • Choose a topic that is suitable for school.<br>  • Examples: a favourite animal, a place of interest, facts about a subject like space, a period in history they are interested in, an important topic for the community. |
| 2 | • Introduce the material on using a search engine (Pages 19 and 20).<br>• Use Discuss 5 (Page 20).<br>• Demonstrate searching using keywords or phrases with the search engine you use.<br>• Ask learners to complete (Page 16) Task B, 'Keywords and phrases', where they list the keywords they will use for researching their topic. | • During the discussion, learners should include keywords such as the name of the team, your local area name, the sport they play and so on.<br>• Talk through the results that you get and how they are related to the keywords.<br>• You could ask learners how they would decide which result to click on. It might be that some are from trusted sources or that they match more of your keywords. It may be that the description is more relevant in one of the results.<br>• Encourage learners to be specific in writing their keywords and phrases. When searching for a topic that is global, they might need to include their region to get results that are relevant to them. |

| Activity | Instructions | Guidance |
|---|---|---|
| 3 | • Tell learners which search engine they will be using and get them to record the information in ( Page 17) Task C, 'Our search engine'.<br>• Explain that they will now be using the search engine to research their topic.<br>• Tell learners how to open the search engine and direct them to complete ( Page 17) Task D, 'Research notes', where they write their research notes. | • You could use a general purpose search engine with 'safe search' turned on. You could also use a search engine on a child-friendly website. |
| 4 | • Introduce the material on searching for images ( Page 20).<br>• Demonstrate searching for images on the search engine you use.<br>• Show learners how to copy an image to their presentation. A quick way is to use right-click and 'Copy image' and then the paste shortcut into the presentation software.<br>• Direct learners to search for and add at least one image to their presentation.<br>• Ask learners to save their work. | • You could use the image search on a general purpose search engine, with safe search turned on, or choose a specific website that has suitable images with permission to reuse.<br>• You could demonstrate how to search for images that you have permission to reuse. On Google, use 'Tools', then 'Usage Rights' and 'Creative Commons' licences.<br>• You may need to remind learners how to switch between the web browser tab with their search engine and their presentation software. |
| 5 | • Remind learners of the importance of staying safe when searching the World Wide Web, whether they are at school, at home or anywhere else.<br>• Use Discuss 6 ( Page 20). | • Reiterate the importance of staying safe when using a search engine outside of school as well as in school. |

**Build on this:**

You could show learners how to copy and paste small pieces of text from websites. Explain that they must give credit to the website. Show learners how to give credit for the images they use.

| Chapter 2: | Lesson 4 Plan and begin your presentation | | |
|---|---|---|---|
| **Learning Objectives** | • Navigate storage hierarchy<br>• Create a presentation with images | | |
| **New key terms** | **resize** – make something bigger or smaller<br>**rotate** – turn something to face a different direction | | |
| **You will need** | • Place the useful images on your local file system or cloud storage. | **Key pedagogy** | Get creative |

| Activity | Instructions | Guidance |
|---|---|---|
| **1** | • Remind learners that they will be continuing with their presentations from the previous lesson.<br>• Demonstrate adding a text box to a presentation slide. Explain that you can hold the mouse down to control the shape of the text box. Demonstrate resizing and rotating the text box and using copy, paste and drag to reposition.<br>• Direct learners to complete Build 4 (Page 21).<br>• Tell learners where to save their work. | • You could show learners how to create a new blank slide or how to select and delete the default items on a new slide.<br>• Learners may accidentally turn a text box upside down when dragging – remind them that they can undo (and what the shortcut is for this). |
| **2** | • Cover the material on inserting images from storage (Pages 21 and 22).<br>• Demonstrate inserting images into a presentation from file storage.<br>• Tell students where they can find the images.<br>• Direct students to add at least one of the useful images to their presentations.<br>• Ask learners to complete (Page 18) Task A, 'Useful images', where they write instructions about how to insert images from storage. | • To insert images from file storage, follow these steps:<br>  • Insert → Image → Upload from computer<br>  • Insert → Image → From drive<br>or similar, depending on your choice of presentation software.<br>• Learners can also continue to search for images. Remind them to stay safe. |
| **3** | • Cover the material on slide designs (Page 22).<br>• Ensure learners understand that each slide should have an image and text.<br>• Ask learners to complete (Pages 18 and 19) Task B, 'Plan your slide designs'.<br>• Learners should start working on their presentation once they have planned a slide. | • So that they don't run out of time, learners should plan one slide at a time before creating it in their presentation software. |

| Activity | Instructions | Guidance |
|---|---|---|
| 4 | • Towards the end of the lesson, remind learners to save their presentations.<br>• Use Discuss 7 (📖 Page 22). Encourage learners to write down tips that they want to use (📖✏ Page 19) Task C, 'Presentation tips'. They can use these tips in the next lesson. | • Learners may suggest:<br>  • making images big enough<br>  • adding more keywords to find just the right image<br>  • dragging and rotating images to make the layout more interesting<br>  • using Ctrl-Z (undo) if you make a mistake. |

**Build on this:**

Show learners how to use the different slide layouts in their presentation software.

| Chapter 2: | Lesson 5 Complete your presentation | | |
|---|---|---|---|
| **Learning Objective** | • Create a presentation with images | | |
| **You will need** | • Saved presentations from the previous lesson | **Key pedagogy** | Get creative |

| Activity | Instructions | Guidance |
|---|---|---|
| 1 | • Introduce the material on text formatting (📖 Page 23). This includes adding image borders, backgrounds and emojis to slides, as well as adding fonts and colours to text.<br>• Demonstrate each technique in your presentation software.<br>• Use Discuss 8 (📖 Page 23). | • Set expectations of the outcomes you expect from learners, based on their previous experience and progress so far.<br>• When leading the discussion encourage learners to think about the people looking at their presentation. Do they like to look at things that are colourful, large and bold, for example? |
| 2 | • Direct learners to open their presentations from the previous lesson.<br>• Ask learners to continue working on their presentations, using (📖✏ Page 20) Task A, 'Finish your slide design' for planning additional formatting.<br>• Remind learners that they need to finish their presentations this week. | • Observe learners' progress and assist any learners who don't have at least one completed slide. |

| Activity | Instructions | Guidance |
|---|---|---|
| **3** | • Ask learners to complete their presentations.<br>• Ask learners to complete ( Page 21) Task B, 'Preparation', where they make notes for what they want to say during their presentation.<br>• Learners who have additional time can complete ( Page 21) Task C, 'Explore', where they explore different features of their presentation software. | • Encourage learners to make notes on things they want to say about their topic so they don't forget.<br>• Task C is optional and can be used for students who work quickly. |
| **4** | • Use Discuss 9 ( Page 23).<br>• Record the tips in a document or on poster paper. | • Some learners may be nervous to present in front of the class. You can offer to support them.<br>• Tips could include:<br>  • speaking clearly<br>  • explaining why they chose each image<br>  • using their Workbook notes to help them remember<br>  • pausing at any point and taking a few seconds to breathe if they are nervous. |
| Build on this:<br><br>Encourage learners to spend time exploring their presentation software. They could make a copy of their presentation before they do this. Remind them how to use 'Undo'. | | |

| Chapter 2: | Lesson 6 Showcase your presentation | | |
|---|---|---|---|
| **Learning Objective** | • Showcase a presentation | | |
| **New key terms** | **slideshow** – full-screen view of slides in presentation software, used when presenting to an audience | | |
| **You will need** | • Saved presentations accessible from a computer connected to a projector | **Key pedagogy** | Showcase |

| Activity | Instructions | Guidance |
|---|---|---|
| 1 | • Remind learners of the presenting tips they suggested in the previous lesson.<br>• Give learners a few minutes to look at their Workbooks and remind themselves of their designs and notes. | |
| 2 | • Demonstrate how to start a full-screen slideshow in your presentation software and how to navigate between slides, (Page 24). | • You need to do the following to start a slideshow and navigate:<br>  • View → Slideshow<br>  • Use right and left arrow keys to move forwards and backwards through the presentation. |
| 3 | • Introduce the showcase to the class.<br>• Invite learners up to the front of the class to give their presentation to their audience. | • Manage time carefully so all learners get a chance to present.<br>• You can run the showcase as a whole class or in smaller groups, if there is not enough time. |
| 4 | • Ask learners to complete (Page 22) Task A, 'Reflection', where learners reflect on the project.<br>• Celebrate the success of your learners and remind them of everything they have learned in this chapter.<br>• Ask 'How will your new skills help you to create interesting documents and presentations? When will you use your new skills?' (Page 25). | • Make links to opportunities for learners to use their skills in other lessons and outside of school, for example, to tell others about a club they go to or to create a book review about a book they enjoyed. |

**Build on this:**

Provide students with an opportunity to give their presentations to another class or group within the school.

# Chapter 3 – Create with code 1

## Project: Create an underwater scene with scenery and two swimming sprites

### Chapter overview

In this chapter, learners will investigate efficiency and repetition before identifying the most efficient algorithms. They will make predictions about animation code and test their theories. Learners will make design choices for their project and add reset code to set the starting points of animated characters. They will experiment and create animations for two sprites that animate at the same time. Learners will test whether their project meets the brief, resets correctly and behaves in the way they intended. They will play the role of tester for other animations and then showcase their animation to the class.

| Chapter 3: Lesson summary | | |
|---|---|---|
| **Lesson** | **Learning Objectives** | **Summary** |
| 1 | • Identify repetition in everyday tasks<br>• Describe efficient algorithms<br>• List the benefits of editing programs to make them clear and concise | Learners complete a word search and discuss their solving methods, including repeated processes. They link this to efficiency and how repetition helps create efficient programs. Learners create their own word search, then complete a word search efficiently. They identify the most efficient algorithms for everyday tasks. |
| 2 | • Follow a linear algorithm and debug to produce the desired output<br>• Predict the outcome of modifications to a sequence<br>• Edit blocks of code to achieve a desired outcome | Learners explore animations and draw costumes for a walking animation based on observations. Learners make predictions about code scripts before testing their predictions in Scratch. |
| 3 | • Develop a program with initialisation code to reset objects to their starting state | Learners start their project for this chapter by making design choices about the backdrop and sprites. They use filtering and searching in Scratch to identify appropriate options. Learners explore the use of reset blocks and identify their purpose in example scripts. They then add their own positional reset blocks to their project. |
| 4 | • Develop a program with more than one object<br>• Develop a program with more than one algorithm running concurrently | Learners begin their project by creating two scripts that run at the same time but run blocks in a different order. They explore student examples, then experiment with blocks in Scratch to plan and write their animation code. Learners describe their animations and discuss their choices. |

| | | |
|---|---|---|
| **5** | • Test a program<br>• Create a program that meets a brief | Learners explore the value of testing projects and complete a testing checklist for their animation. They get two other people to test their animation and confirm it runs as described, resets and meets the brief. |
| **6** | • Share a program | Learners take part in a Scratch swimming gala to showcase their swimming animations. They then reflect on their skills and celebrate all they have achieved. |

| End of chapter project | **Create an underwater scene with scenery and two swimming sprites** |
|---|---|
| **Example ideas that learners could design:** |    |

## Chapter 3 – Create with code 1

### Project: Plan a journey for a Bee-Bot to take to tell a story

| | |
|---|---|
| **Name** | |

| | **Working towards (1)** | **Meets expected (2)** | **Exceeding expected (3)** | **Score** |
|---|---|---|---|---|
| **Sketches** | • Created Scratch project does not have scenery sprites and/or a backdrop | • Created Scratch project has scenery sprites and a backdrop | • Created Scratch project has scenery sprites and a backdrop. Reset blocks have been added to the scenery incase they are accidentally moved | |
| **Final design** | • Swimming sprites are animated but do not reset when the flag is clicked. | • Swimming sprites are animated and reset when the flag is clicked. | • Swimming sprites are animated and reset when the flag is clicked. Code is commented. | |
| **Everyday task** | • The project meets some items in the testing checklist | • The project meets all items in the testing checklist and has been tested by someone else | • The project meets all items in the testing checklist and has been tested by someone else. Feedback and improvements from observations have been added | |

| | |
|---|---|
| **Teacher feedback** | |
| **Learner response** | |

| Chapter 3: | Lesson 1 Word search patterns |
|---|---|
| **Learning Objectives** | • Identify repetition in everyday tasks<br>• Describe efficient algorithms<br>• List the benefits of editing programs to make them clear and concise |
| **New key terms** | **efficient** – do something in the quickest and most direct way<br>**repetition** – do the same thing again and again |
| **Key pedagogy** | Classroom talk |

| Activity | Instructions | Guidance |
|---|---|---|
| **1** | • Remind learners that they already know that algorithms are precise sets of instructions and that they have previously used the 'repeat' code block. | • The 'What do you already know?' bullet points both relate to Stage 2 learning. This is when they used a Bee-Bot to follow a precise set of instructions and create a Scratch project with repeated movements. |
| **2** | • Introduce word searches (Page 27).<br>• Ask learners to complete (Page 23) Task A, 'Complete a word search'.<br>    • Learners complete a word search. | • You could ask learners if they have done a word search before. They may have already completed one in school or at home.<br>• Point out that sometimes word searches are only horizontal and vertical, but in some words can be diagonal too. |
| **3** | • Use Discuss 1 (Page 27). | • Encourage learners to express different ways of solving a word search. There isn't one right answer, but there are ways that are more efficient than others.<br>• Remember answers that reference methods with efficiency or repetition as these will be useful later. |

| Activity | Instructions | Guidance |
|---|---|---|
| 4 | • Talk through repetition (Page 27). Reference the repetition work learners did in Stage 2, 'Create with code'.<br>• Use Discuss 2 (Page 27).<br>• Follow up with the list of examples.<br>• Explain that you can repeat something a set number of times, or you can repeat until something happens. | • Prompt learners to think about the examples they gave when answering the first discussion question. Can they spot any repetitive parts to completing a word search?<br>• There are some examples on page 27 of the Student's Book but they are not exhaustive.<br>• You could pull up the word search from the Workbook on the board and demonstrate each example. |
| 5 | • Talk through efficiency (Page 28). Make reference to Stages 1 and 2, 'Our digital world', where learners discovered that robots can complete some tasks more efficiently than humans.<br>• Show learners the pictures of the code blocks (Page 28). Explain the benefits of efficient algorithms.<br>• Discuss 3 (Page 28).<br>• Talk through the example efficient tips for completing a word search. | • Ask learners to count the number of blocks in the script with no repeats and then the script with repeats. Which is written more efficiently for a human to read?<br>• Learners might identify the tips in the Student's Book. There are others such as:<br>  • looking for double letters as they stand out more. In their word search, there are words with 'dd', 'zz', 'tt' and 'll'<br>  • tracing where you are up to with your finger or blocking off rows using a piece of paper. This helps you keep focus and be more efficient. |
| 6 | • Ask learners to complete (Page 24) Task B, 'Make a word search'.<br>  • Learners create their own word searches.<br>• Tell learners to swap their word searches with a partner and complete them as efficiently as they can.<br>• Ask 'Did following your class tips help you complete your word search efficiently?' (Page 28). | • Learners should choose nine words for their word search. You can link this to another topic they are covering, or you could ask them to choose words linked to computing or their hobbies or interests. |

| Activity | Instructions | Guidance |
|---|---|---|
| 7 | • Ask learners to complete (Page 25) Task C, 'Efficient algorithms'.<br>    • Learners choose the most efficiently written algorithm from a number of everyday algorithms.<br>• Discuss the answers. | • You could demonstrate the two keyboard-related algorithms and refer to Chapter 2.<br>• You could show videos for braiding hair and signing 'dog'. Note that this is British Sign Language and other sign languages will vary. |
| **Build on this:**<br>Ask learners to write an algorithm for solving a word search. | | |

| Chapter 3: | Lesson 2 Costume change | | |
|---|---|---|---|
| **Learning Objectives** | • Follow a linear algorithm and debug to produce the desired output<br>• Predict the outcome of modifications to a sequence<br>• Edit blocks of code to achieve a desired outcome | | |
| **New key terms** | **animation** – making characters and objects look like they are moving | | |
| **You will need** | • The starter Scratch project (Website C3L2R1) | **Key pedagogy** | Develop practical skills |

| Activity | Instructions | Guidance |
|---|---|---|
| 1 | • Remind learners of the terms 'predict' and 'debugging'. | • Learners were first introduced to 'predict' and 'debugging' in Stage 1 with Bee-Bot paths and Scratch layers. |
| 2 | • Introduce the concept of animation and talk through the examples of animation code (Page 29). | • Use the costume dragging screenshot to highlight that the walking animation would not look like a natural walk if the costumes were in the wrong order. |
| 3 | • Split learners into pairs and ask them to complete (Page 26) Task A, 'Walking poses'.<br>    • Learners watch each other walking before drawing their repeated movements in order.<br>• Use Discuss 4 (Page 29). | • You might need to ask learners to slow down the walk so that the poses can be detected.<br>• If you prefer, you could play videos of people walking slowly instead of asking learners to act out a walk. |

| | | |
|---|---|---|
| **4** | • Talk through the examples of animation code (📖 Page 30). | • You could ask a learner to demonstrate their usual walk, then to do it with a one or two second delay between poses. Which speed do learners think is about right for a walk?<br>• Ask learners if they think the beetle will spin quickly or slowly, based on the code on page 30 of the Student's Book. |
| **5** | • Ask learners to complete (📖✏ Pages 26 and 27) Task B, 'Keep Manpreet warm'.<br>  • Learners will need the starter Scratch project (Website C3L2R2).<br>  • Learners make predictions about the code blocks they are given.<br>• Discuss the answers. | • Learners can work in pairs and discuss their predictions. Talking through code out loud is a good skill to boost confidence, understanding and communication.<br>• Make sure learners are editing their code rather than creating multiple new scripts that each run when they click the green flag. |
| **6** | • Ask 'Were the predictions that you made about the Manpreet sprite code blocks correct? Did anything surprise you about the animation? You were also asked to edit the code. How did you make changes to the blocks of code?' (📖 Page 30). | • Learners have seen animation with costumes, movement and rotation.<br>• They have seen animation that names specific costumes and those that move to the next in order.<br>• They have seen fast and slow animations using 'wait' blocks. |

Build on this:

Learners can experiment with animations in Scratch by using their 'Keep Manpreet warm' starter project. For example, they can drag in different motion and looks blocks; they can change the numbers in blocks such as repeat, wait and steps to see the impact.

| **Chapter 3:** | **Lesson 3 Project design** | | |
|---|---|---|---|
| **Learning Objective** | • Develop a program with initialisation code to reset objects to their starting state | | |
| **New key terms** | **reset** – put objects back to their starting point in Scratch | | |
| **You will need** | • Computers that can access Scratch<br>• Scratch accounts or another way to save learners' Scratch projects | **Key pedagogy** | Get creative |

| Activity | Instructions | Guidance |
| --- | --- | --- |
| **1** | • Re-introduce the project using the project brief and examples (📖 Page 31). | • The most important thing to get across to learners is that not all sprites need to be animated, but they must have two sprites animated at the same time. |
| **2** | • Get learners on to computers with a new Scratch project.<br>• Ask them to complete Build 1 (📖 Page 31) to choose a backdrop. | • Remind learners to log on to their Scratch account or practise saving to the computers. It is important they can access their saved Scratch project in the next lesson.<br>• Users should give their project a sensible name so they can find it again easily.<br>• Tell learners that they can click on the 'Underwater' button to see only underwater backdrops.<br>• There is a 'pool' backdrop that can also be used, but learners will need to search/scroll for this as it does not appear in the underwater category. |
| **3** | • Ask learners to complete Build 2 (📖 Page 31) to choose scenery sprites.<br>• Use Discuss 5 (📖 Page 31). | • Learners can have more than one of the same scenery sprite, for example if they want to create a pile of rocks.<br>• Remind learners that they can drag a sprite on the stage to put it in front of other sprites (this was covered in Stage 1 when learners created a layered window scene).<br>• You could demonstrate the example from the Student's Book showing how to edit the fishbowl costume to create reed scenery. Can learners find any other useful edits? |

| Activity | Instructions | Guidance |
|---|---|---|
| 4 | • Ask learners to complete Build 2 (Page 32) to choose swimming sprites. | • Learners should not animate the swimming sprites yet; just position them where they want the animation to start.<br>• If learners want to create a costume changing animation, they should look for a sprite with multiple costumes or be prepared to add or duplicate and edit costumes.<br>• The keyword searches are useful for narrowing down results in Scratch (and make a great link to the previous chapter), but there are some issues. The shark2 sprite is one example as it doesn't appear in the 'underwater' search. |
| 5 | • Ask learners to complete (Page 28) Task A, 'Project design checklist'.<br>  • Learners work through a checklist to ensure they have completed all the build tasks from the Student's Book correctly. | • The checklist will ensure learners have all the elements needed to meet the project brief.<br>• You could split learners into groups so they can talk about their design choices and reasons. |
| 6 | • Talk through the reset explanation text and the annotated example (Page 32).<br>• Use Discuss 6 (Page 32).<br>• Ask learners to complete (Page 29) Task B, 'Sprite reset'.<br>  • Learners answer questions about reset blocks of code in Scratch. Discuss the answers. | • Encourage learners to think of computer games that they play where reset blocks are used. They might know a position reset as a 'respawn'. Some games have costume resets to basic costumes or reset storage chests to empty. Others might reset the size of a character or the direction they face.<br>• When discussing the Workbook answers, you could demonstrate the code with and without the reset blocks by recreating it in a new Scratch project. |

| Activity | Instructions | Guidance |
|---|---|---|
| **7** | • Ask learners to complete Build 4 (📖 Page 32) to reset the position of their swimming sprites.<br>• Learners should test that the reset block works as expected. | • If the reset block does not reset the swimming sprite to the correct position, learners should delete the 'go to x y' block. Then they should drag the sprite to the starting position and drag the block to the code area again. This is because the block will be filled with the x/y coordinates of the sprite at the time the block was added.<br>• The scenery will not be draggable in full-screen mode, but when they are testing their animations in the editor during the next few lessons they may accidentally move the scenery. |

**Build on this:**

Learners could add reset blocks to their scenery sprites to practise their dragging and connecting of blocks.

| Chapter 3: | Lesson 4 Swim time | | |
|---|---|---|---|
| **Learning Objectives** | • Develop a program with more than one object<br>• Develop a program with more than one algorithm running concurrently | | |
| **New key terms** | **script** – collection of Scratch blocks that are attached to one another; a sprite may have many scripts | | |
| **You will need** | • Learners' Scratch underwater animation projects from the previous lesson | **Key pedagogy** | Develop practical skills |

| Activity | Instructions | Guidance |
| --- | --- | --- |
| **1** | • Set up learners on the computers with their projects from the previous lesson.<br>• Ask learners to complete Builds 5 and 6 ( Page 33), using the say and wait blocks.<br>• Ask learners to complete Build 7 ( Page 33), testing their code by clicking on the green flag. | • Highlight that the code for both sprites will run at the same time, but they will see the bubbles at different times because of the wait block.<br>• Learners might need help using the shortcut to get to the emoji keyboard. Remind them that they used shortcut keys in the 'Content creation' chapter. |
| **2** | • Show learners the student examples of animation code ( Page 34).<br>• Ask learners to complete Build 8 ( Page 34). At the same time, they should complete ( Page 30) Task A, 'Animation planning', where they note down any blocks they want to use in their project. | • Encourage learners to make note of code blocks in the student examples that they would like to try.<br>• They can use the planning space in the Workbook to make notes of things they would like to try, and then cross them out if they don't want to use them. |
| **3** | • Ask learners to complete ( Page 31) Task B, 'My animation code'.<br>  • Learners describe what their animation does.<br>• Use Discuss 7 and Discuss 8 ( Page 34). | • Learners could use their Workbooks to explain their code to someone else and discuss similarities and differences. |

**Build on this:**

Hold a class survey about the code blocks used. Ask the first learner to list their blocks, one at a time. Everyone who used that block should put up their hands. Which blocks were most popular? Were any blocks used only once?

| **Chapter 3:** | **Lesson 5 Meet the brief** | | |
| --- | --- | --- | --- |
| **Learning Objectives** | • Test a program<br>• Create a program that meets a brief | | |
| **New key terms** | **testing** – trying a project out to make sure it does everything it is supposed to do | | |
| **You will need** | • Learners' Scratch underwater animation projects from the previous lesson | **Key pedagogy** | Real-world contexts |

| Activity | Instructions | Guidance |
|---|---|---|
| 1 | • Read through the testing recap (Page 35). Include a reminder of the tests that learners have carried out so far in this chapter and the benefits of testing.<br>• Explain that they might need extra reset blocks, but this will depend on each learner's animation. | • Learners can add their reset blocks to their reset scripts created in Lesson 3. These will already have a 'when flag clicked' block and a 'go to x y' block. |
| 2 | • Ask learners to complete ( Page 32) Task A, 'Testing checklist'.<br>  • Learners complete a checklist to ensure that their project meets the brief. | • Learners should test their animation multiple times and mark each checkbox once they are happy that they have met that part of the brief.<br>• This is a time for learners to fix any errors. Encourage them to help each other carry out fixes.<br>• There is also time here to make any enhancements. The previous lesson ended with discussions about the various code blocks and favourite parts. Learners will have gained inspiration from their classmates that they may want to add to their project. |
| 3 | • Ask learners to complete ( Page 33) Task B, 'Testing confirmation'.<br>  • Learners work in groups of three to test their projects.<br>• Use Discuss 9 ( Page 35). | • Learners should choose two people to test their project and confirm that it meets the brief.<br>• You could encourage learners to choose people with projects that appear very different from theirs.<br>• If learners had difficulty with a particular block, they could specifically ask for someone who has used that block to test their animation. |

**Build on this:**

You could be an additional tester for each learner and give your 'official' testing stamp of approval to the animations.

| Chapter 3: | Lesson 6 Swimming gala | | |
|---|---|---|---|
| **Learning Objective** | • Share a program | | |
| **You will need** | • Learners will need their Scratch projects | **Key pedagogy** | Showcase |

| Activity | Instructions | Guidance |
|---|---|---|
| **1** | • Hold a 'swimming gala' – an event where learners showcase their swimming animations.<br>• Give learners time to practise their showcases so they are confident with their animations.<br>• Ask learners to showcase their animations in full-screen mode. Decide how to run the showcase with your class to ensure that each learner gets to present. This could be in groups or pairs, if needed. | • Ask learners questions about their reasons for their design choices for backdrop and sprites.<br>• Ask learners what they needed to reset in their animations.<br>• Ask learners if they found anything in testing that they needed to fix or wanted to improve. |
| **2** | • Ask learners to complete (Page 34) Task A, 'Gala medal', where they decorate a gala medal to celebrate their success in the project.<br>• Ask learners to complete (Page 35) Task B, 'Reflection', where they reflect on the project. | • Encourage learners to enjoy these two Workbook tasks by taking time to celebrate all that they have achieved during this chapter. |
| **3** | • Read through the celebration page (Page 37) to recap the chapter.<br>• Ask 'How important do you think it is to test your project? Why do you need to test your project more than once?' (Page 37). | • Learners might answer that they found out things about their animation when they played it all together right through. It may have given them an opportunity to make changes and fix errors.<br>• They might mention that it was an important step to make sure that the animation met the project brief.<br>• Learners should comment that testing more than once is a good way to make sure their sprites reset correctly. |

| Activity | Instructions | Guidance |
| --- | --- | --- |
| **Build on this:** Create a class studio in Scratch showcasing learners' swimming animations, which can be shared with others in the school community. Give learners time to play the projects in the studio and 'see inside' any projects that they want to explore further. | | |

# Chapter 4 – How computers work

## Project: Design an Internet of Things device to help reach a UN Sustainable Development Goal

## Chapter overview

In this chapter, learners will label and sort hardware and software. They will investigate various manual and automatic input devices and explore the benefits of robots in manufacturing. They will investigate the Internet of Things and design a device to help reach a United Nations Sustainable Development Goal.

| Chapter 4: Lesson summary | | |
|---|---|---|
| **Lesson** | **Learning Objectives** | **Summary** |
| **1** | • Define hardware and software and their roles in a computer system<br>• Describe how hardware and software work together to form a working system | Learners explore the roles of different hardware and software. They act out scenarios where hardware and software work together to complete tasks. |
| **2** | • List a range of manual and automatic input devices | Learners explore the roles of manual input devices and identify automatic input devices. They then craft a night light with an embedded, pre-coded BBC micro:bit device. |
| **3** | • Describe how robots are used in manufacturing | Learners discover how robots are used in manufacturing. They then complete a role-play activity of a production line to explore efficiency and accuracy. |
| **4** | • Explain how computers can be programmed to control other machines and objects<br>• List common Internet of Things (IoT) devices in the world around us | Learners define a computer-controlled device and identify them in their school environment. They then explore the term Internet of Things and create a design for an IoT device that could be used in school. |
| **5** | • Design an IoT device to support a UN Sustainable Development Goal | Learners explore the UN Sustainable Development Goals. They sketch several ideas for an Internet of Things device that they could make to support one of the UN Sustainable Development Goals. They begin sketching their final design. |
| **6** | • Showcase the design of an IoT device | Learners finish their IoT designs and practise presenting their ideas. They then present their ideas to the class before reflecting on the project. |

| End of chapter project | **Design an IoT device to help reach a UN Sustainable Development Goal** |
|---|---|
| **Example ideas that learners could design:** |  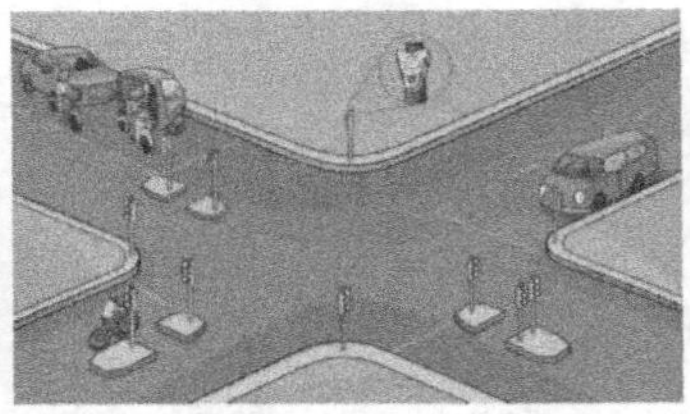  |

## Chapter 4 – How computers work

### Project: Make a counting app to help with counting from 1 to 5

| **Name** | |
|---|---|

| | **Working towards (1)** | **Meets expected (2)** | **Exceeding expected (3)** | **Score** |
|---|---|---|---|---|
| **Sketches** | • Minimal design work had been completed for the initial design | • Three design sketches had been created with some explanations of what the device would help with | • Three design sketches had been created with detailed explanations of what the device would help with | |
| **Final design** | • Minimal design work had been completed for the final, chosen design | • A final, labelled design had been created | • A final, detailed design had been created that was clearly labelled | |
| **Everyday task** | • A minimal attempt had been made to reflect on the final design | • The learner described what made their device helpful and how it could be improved | • The learner used a high level of detail in their reflection to describe two of the most helpful parts and how the device could be improved | |

| **Teacher feedback** | |
|---|---|
| **Learner response** | |

| Chapter 4: | Lesson 1 Hardware and software | | |
|---|---|---|---|
| **Learning Objectives** | • Define hardware and software and their roles in a computer system<br>• Describe how hardware and software work together to form a working system | | |
| **You will need** | • Scrap paper and pens | **Key pedagogy** | Unplugged |

| Activity | Instructions | Guidance |
|---|---|---|
| **1** | • Introduce the project that learners are working towards (Page 38).<br>• Talk through the example projects:<br>  • a smart watering sprinkler that can be switched on and off by a farmer from their phone wherever they are<br>  • a smart traffic device that monitors the number of cars on the road and changes traffic signals<br>  • a smart bin device that senses when it is full; it then sends a message that it needs to be emptied so that it doesn't overflow. | • As you talk through the illustrations, you could ask learners why the devices would help make a better world. Here are some suggestions for each device:<br>  • The smart sprinkler helps save water.<br>  • The smart traffic device helps keep traffic flowing and, as a result, reduces air pollution.<br>  • The smart bin stops the bin overflowing and littering a public area. |
| **2** | • Talk through the images of hardware and the role of each (Page 39). Draw attention to any hardware in the classroom.<br>• Use Discuss 1 (Page 39). | • Learners should be familiar with the names of the hardware covered in this lesson. |
| **3** | • Talk through the images of software and the role of each (Page 40). Refer to occasions when learners have used the software in lessons.<br>• Use Discuss 2 (Page 40). | • Learners should be familiar with the names of the software covered in this lesson from Stages 1 and 2, or from using the software outside of school. |
| **4** | • Ask learners to complete (Page 36) Task A, 'Hardware and software'.<br>  • Learners choose the best pair of hardware and software for different scenarios. | • If there is time, you could ask learners to peer-mark each other's work. |

| Activity | Instructions | Guidance |
| --- | --- | --- |
| **5** | • Look through the illustrations of hardware and software working together (Page 41).<br>• Introduce the role-play activity (Page 41). Learners will be acting out examples in which hardware and software work together to make something happen.<br>• Split the class into groups of three learners.<br>• Ask the groups to choose one example from the role-play activity. They should assign each group member to either a piece of hardware or a piece of software.<br>• Give learners time to plan and practise their examples – encourage creativity and have fun!<br>• At the end of the time, learners should take turns to act out their example. Ask the class if they can guess which scenario is being acted out.<br>• Learners should now have a clearer idea of how hardware and software work together. | • Ask learners to think creatively about how they could act out a specific role. They could act out or speak the action, or make use of props such as paper to create basic icons and buttons.<br>• An example might be: Mouse → presentation software → printer. Using pieces of paper, one group creates a drawing of a cursor arrow, a print icon and a presentation slide. The 'mouse' moves with the cursor and clicks the print icon in the 'presentation software'. At a distance, the 'printer' acts out the printing action by producing a piece of paper with the slide on. |
| **6** | • Ask learners to complete (Page 37) Task B, 'Hardware and software work together'.<br>    • Learners reflect on the role-play activity. You can discuss the questions as a group. | • Learners might have had difficulty thinking about how their device interacted with the next device in the role play. Refer to wired and wireless connections from Stage 1.<br>• Learners might not have been confident that they knew the role of their device. Refer to the descriptions of hardware and software from earlier in the lesson to reinforce their roles. |

**Build on this:**
Continuing in their groups, learners could think of their own scenarios of using familiar hardware and software to 'act out'.

| Chapter 4: | Lesson 2 Input devices | | |
|---|---|---|---|
| **Learning Objective** | • List a range of manual and automatic input devices | | |
| **New key terms** | **manual input device** – device that needs a human to input (enter) data; for example, you press the keys on a keyboard to enter words and numbers into the computer<br>**automatic input device** – device that collects data without needing a human to input (enter) data; for example, a temperature sensor collects the temperature from the air around it or a barcode reader reads data from a barcode<br>**micro:bit** – tiny computer that has inputs and outputs that you can program to make fun projects | | |
| **You will need** | • micro:bits (enough for paired work)<br>• micro:bits should be pre-programmed with the code in Step 2 on the micro:bit website (Website C4L2R1). See the hardware and software guidance at the start of the Teacher's Guide for how to pre-program.<br>• Crafting materials | **Key pedagogy** | Physical computing |

| Activity | Instructions | Guidance |
|---|---|---|
| **1** | • Talk through the images of the manual input devices (Page 42).<br>• Use Discuss 3 (Page 42).<br>• Ask learners to complete (Page 38) Task A, 'Manual input device roles'.<br>  • Learners match the manual input device to its role. | • If there are devices that they have not used in school, you could discuss the answers as a class and draw particular attention to those device roles. |
| **2** | • Talk through the images of the automatic input devices (Page 43).<br>• Use Discuss 4 (Page 43).<br>• Ask learners to complete (Page 38) Task B, 'Identify the automatic input device'.<br>  • Learners circle the automatic input devices. | • Though these devices could be found in school settings, it is unlikely that your school will have all the devices.<br>• For any devices not seen in school, you could mention an example from the local community (for example, in shops or at automated cash machines). You can talk about your experience with these devices either within the school or the community to make them more relevant. |

| | | |
|---|---|---|
| **3** | • Before the lesson, pre-program enough micro:bits (Website C4L2R2) for learners to share (see 'You will need' for guidance).<br>• Introduce the material about micro:bits and the night light task (Page 44).<br>• Use Discuss 5 (Page 44).<br>• Explain that a night light comes on automatically when it is dark, in response to a light sensor.<br>• Show the class the micro:bits and give a short demonstration showing how the lights turn on when the room gets dark. You can place your thumb over the sensor to test.<br>• Ask learners to complete (Page 39) Task C, 'Design your night light'.<br>  • Learners design their night lights. | • If you have a night light that turns on when it gets dark, you could bring it in to show learners.<br>• Tell learners which crafting materials they will have access to (e.g. paper, felt pens, cardboard).<br>• Walk around the room and observe learners creating their designs. Give inspiration to any learners who are struggling to make a design choice. They could use their hobbies, interests or favourite items.<br>• Explain that learners should not cover the LEDs in their design so that the sensor can react to the light level. |
| **4** | • Ask learners to begin crafting their night lights.<br>• As the lesson reaches the end give learners time to share their designs with another pair or with the class. | • As learners build and craft their night lights walk around the room and ask questions about the automatic light sensor. Ask them about the difference between automatic and manual inputs. |

**Build on this:**

Learners could provide feedback for each other and improve their designs based on this feedback.

| **Chapter 4:** | **Lesson 3 Robots** | | |
|---|---|---|---|
| **Learning Objective** | • Describe how robots are used in manufacturing | | |
| **You will need** | • Materials prepared for the role-play activity | **Key pedagogy** | Role play |

| Activity | Instructions | Guidance |
|---|---|---|
| **1** | • Recap learners' knowledge about robots and introduce robots in manufacturing (Page 45). | • Learners were first introduced to robots in Stage 1 , 'Our digital world'. Remind them that they looked at robots in the real world. In their project they designed a robot to help someone at school.<br>• Learners may be most familiar with robots that look like humans. Make sure they understand that robots come in many shapes. This was covered in Stage 2, 'Our digital world'. |
| **2** | • Introduce the role-play activity (Page 46).<br>• This activity involves each learner playing the part of a robot on a production line. Decide what object your class is going to build on their production line. It could be:<br>  – a bird made from a few plastic building blocks<br>  – an origami animal<br>  – icing a gingerbread man.<br>• Tell the class that they will be playing the part of a robot in a production line.<br>• Ask the class to observe as you demonstrate the production line in full (this will depend on your chosen materials).<br>• Split the class into small groups and give them the materials needed to manufacture the item.<br>• The groups should decide which task each 'robot' will perform (for example, one origami fold, adding one plastic building block, or decorating one icing feature).<br>• Give them time to practise the line slowly before coming together to quickly complete the manufacturing process in full. | • Observe the groups and provide support where needed. For example:<br>  – Can you do your task any faster?<br>  – Are you making any errors?<br>  – Are you enjoying your task?<br>  – Is it easier just focusing on one part?<br>• Remind learners of their roles and complete additional demonstrations where needed. |

| Activity | Instructions | Guidance |
|---|---|---|
| **3** | • Ask learners to complete (Page 40) Task A, 'Reflection'.<br>  • Learners reflect on the role of robots in manufacturing and the role-play activity.<br>• Hold a class vote: 'Who thinks robots are better than humans in manufacturing?'<br>• Explain that robots are better for some tasks but humans are needed to design robots and decide what they should do (Page 46). | • Check learners' reflection statements and monitor participation in discussion.<br>• Questions 1 and 2 from the reflection activity are expected to be answered from the lists on pages 45 and 46 of the Student's Book.<br>• Answers for question 4 could include: 'Yes, because they could be quicker or more accurate', or 'No, because when something went wrong they couldn't fix it as quickly.' |

**Build on this:**
After some practice time, groups could race to produce as many objects as possible within a time limit. The winning group will have the most objects made, and each object must be of good quality to pass the test. Discuss how the introduction of time pressure impacted the quality and number of errors. Would a robot be able to speed up without loss of quality? Did human emotions brought about through competitiveness play a part in accuracy?

| Chapter 4: | Lesson 4 Internet of Things | | |
|---|---|---|---|
| **Learning Objectives** | • Explain how computers can be programmed to control other machines and objects<br>• List common Internet of Things devices in the world around us | | |
| **New key terms** | **computer-controlled device** – device that performs a task based on inputs<br>**Internet of Things** – network of everyday devices that are connected through the internet | | |
| **You will need** | • Identify some potential Internet of Things devices in the school environment to reference in the lesson. | **Key pedagogy** | Get creative |

| Activity | Instructions | Guidance |
| --- | --- | --- |
| **1** | • Introduce the concept of computer-controlled devices (Page 47).<br>• Use Discuss 6 (Page 47). | • Provide feedback on the devices that learners have identified in the classroom. Are they likely to be computer controlled? You may not be certain for some devices. That's fine, just explain that we can't always tell.<br>• Examples from the classroom might include:<br>    ◦ electronic heating or air conditioning controls<br>    ◦ security alarm system.<br>• Remind students that a simple (on/off) electronic device (such as a torch) would not be a computer-controlled device. |
| **2** | • Introduce the concept of the Internet of Things (IoT) (Page 48).<br>• Use Discuss 7 (Page 48).<br>• Use Discuss 8 (Page 48). | • If your learners are stuck for ideas, revisit the examples in the Student's Book.<br>• Further examples include: a payment device in a supermarket, a fitness tracker, a security alarm, a video doorbell and a bird-watching webcam that you can view on the internet. Draw on local examples that your learners will be familiar with.<br>• A potential answer to discussion question 8 is an air conditioning unit. If it was controlled through the internet, the headteacher could switch the air conditioning on or off before they came to school. |

| Activity | Instructions | Guidance |
|---|---|---|
| 3 | • Split learners into pairs.<br>• Ask learners to complete (Page 41) Task A, 'Design an Internet of Things device to help your school'.<br>  • Learners design an IoT device for their school. Each learner in the pair will draw the design in their own Workbook. | • Use the examples (Page 48) to help guide learners who might be struggling to develop an idea.<br>• Additional ideas:<br>  • Could any supplies be ordered when they run out?<br>  • Is there somewhere in the school that a remote-controlled webcam would be helpful?<br>  • Could something be controlled based on the weather or time of day? |
| 4 | • Ask pairs to present their IoT device ideas to the class (see 'Showcase', (Page 48). | • Check that learners have designed an IoT device that would support the school. Their explanation should show *how* it benefits from being connected to the internet.<br>• To support learners, you could ask these questions:<br>  • What does your device do?<br>  • Why does your device need to be connected to the internet?<br>  • Why is your device better by being connected to the internet? |

**Build on this:**
Learners could use any feedback that they had from their showcase to inform the design of their IoT device in the next lesson.

| Chapter 4: | Lesson 5 Design your project | | |
|---|---|---|---|
| **Learning Objective** | • Design an IoT device to support a UN Sustainable Development Goal | | |
| **You will need** | • Coloured pencils or pens to support learners with their designs<br>• **Optional:** 'World's Largest Lesson Animation: Part 1' video (go to the World's Largest Lesson website, Resources section and search for Malala) | **Key pedagogy** | Get creative |

| Activity | Instructions | Guidance |
| --- | --- | --- |
| 1 | • If possible, use Malala Yousafzai's video introduction to the global goals from the World's Largest Lesson website to introduce the UN Sustainable Development Goals.<br>• Reinforce what was introduced in the video (Page 49).<br>• Use Discuss 9 (Page 49). | • Check learners' understanding of the purpose of the UN Sustainable Development Goals.<br>• Draw attention to the goal or goals that are most relevant to your setting. These could be Good Health, Quality Education, Climate Action or any other goals that are relevant. |
| 2 | • Introduce the project brief (Page 50).<br>• Narrow down the 17 goals to a few goals most relevant for your setting.<br>• Ask the class to work in pairs to design a device to support a goal (Pages 42, 43 and 44) Task A, 'Design an Internet of Things device to support a UN Sustainable Development Goal'.<br>• Explain that learners will have the rest of this lesson and the first part of the next lesson to draw, label and colour their designs.<br>• Tell them that they will be sharing their designs with the class so they should complete the Workbook questions to help them describe their device. | • Walk around the room and observe the progress of the pairs. Provide support and guidance as needed.<br>• Encourage learners to label their designs as this will help when they are showcasing them later.<br>• If learners are stuck for ideas, they should revisit the examples on page 50 of the Student's Book. |

**Build on this:**

Learners could ask for feedback from peers part way through their designs and use this to help them improve their ideas. Feedback could include improvements to the device design or how it connects to the internet. It might also include observations from a user's point of view or potential additional benefits the learner hadn't thought about.

It would also be good for learners to think about questions asked by their peers, as this will highlight areas in their drawing, labelling or explanation that could be strengthened before the showcase.

| Chapter 4: | Lesson 6 Showcase your Internet of Things device | | |
|---|---|---|---|
| **Learning Objective** | • Showcase the design of an Internet of Things device | | |
| **You will need** | • Coloured pencils or pens to support learners with their designs<br>• **Optional:** (Website C4L6R1) (go to the World's Largest Lesson website, Resources section and search for certificate), printed for each student | **Key pedagogy** | Showcase |

| Activity | Instructions | Guidance |
|---|---|---|
| **1** | • Remind learners that they started to design their Internet of Things devices in pairs during the previous lesson.<br>• Give learners some time to finish their designs. | • Remind learners to label their designs to help them describe what their device does.<br>• Some learners may need help explaining how their device uses the internet.<br>• Some may need help explaining their idea clearly to other learners. |
| **2** | • Ask pairs to practise presenting their Internet of Things device to another pair. | • They should ask for feedback on how well their presentation explains their idea.<br>• Remind learners that they will be presenting the designs from their Workbooks and they can use the labels they added to guide their showcase. |
| **3** | • Ask pairs to present their Internet of Things device to the class.<br>• Encourage other learners to think about how the devices would help your school. | • You can ask learners to showcase in small groups or as a class, depending on the class size and the time available.<br>• Prompt learners to expand on how the device works and how it helps meet their goal. |

 

| Activity | Instructions | Guidance |
|---|---|---|
| 4 | • Ask learners to complete (Page 45) Task A, 'Reflection'.<br>    • Learners reflect on their projects.<br>• You could ask learners to share some of their thoughts with the rest of the class.<br>• You could end the chapter by giving out 'World's Largest Lesson' global goal participation certificates to the class. | • Read the reflection questions with learners to make sure they know what they have been asked to do.<br>• You might want to keep a log of the feedback that learners receive to help inform their future work. |

**Build on this:**
Ask pairs to peer review the idea presented by another pair. They should ask for two good things and one wish to make it better. Pairs could improve their designs based on feedback.

# Chapter 5 – Create with code 2

## Project: Design and build a project that includes one input and one output, using a BBC micro:bit

### Chapter overview

In this chapter, learners will begin with an unplugged lesson where they create sets of dance moves and edit them until they are happy with their performance. They will then be introduced to micro:bits and the MakeCode platform. They will identify inputs and outputs and build programs from given blocks that produce a desired output. They will then create their own projects in pairs that include one input and one output.

Note: Sound is built in to the micro:bit v2. If you have a micro:bit v1, you will need to add a speaker to enable sound. You can also use the micro:bit simulator throughout, if you do not have access to micro:bit devices.

| Chapter 5: Lesson summary | | |
|---|---|---|
| **Lesson** | **Learning Objectives** | **Summary** |
| 1 | • Describe logical thinking and its purpose<br>• Describe how tasks can be divided into smaller sections to make them easier to follow and modify | Learners describe how logical thinking helps to follow and create algorithms. They work in groups to create an algorithm for a dance by breaking it down into smaller parts. They edit their algorithm and recognise that it is easy to work with smaller parts of an algorithm. |
| 2 | • Identify the inputs of algorithms<br>• Develop a sequence to produce an output based on an input | Learners identify the inputs and outputs used in a block-based program for the micro:bit. They predict the outputs that will be produced by programs and then enter and run the code and discuss the results. |
| 3 | • Describe how to test and debug programs so that they produce the desired output<br>• Explain how programmers use their mistakes to inform the programs that they create | Learners recall the importance of testing their programs. They test a program they create and then make changes. They recognise and fix common bugs that occur when using the micro:bit and contribute to a class list of bugs and fixes. |
| 4 | • Create a program that produces an output from an input device | Learners are introduced to the project brief. In pairs, they design an algorithm for a micro:bit project that uses at least one input and one output. |
| 5 | • Develop a program for a physical computing device that produces an output | Learners work in pairs to create the code for their algorithm and then test and debug their programs. |

| | | |
|---|---|---|
| **6** | • List the benefits of collaboration in programming | Learners showcase their projects to other pairs. They discuss the benefits of working collaboratively when programming. They share their experience of testing and debugging with the class. |

| End of chapter project | **Design and build a project that includes one input and one output, using a BBC micro:bit** | | |
|---|---|---|---|
| **Example ideas that learners could design:** | 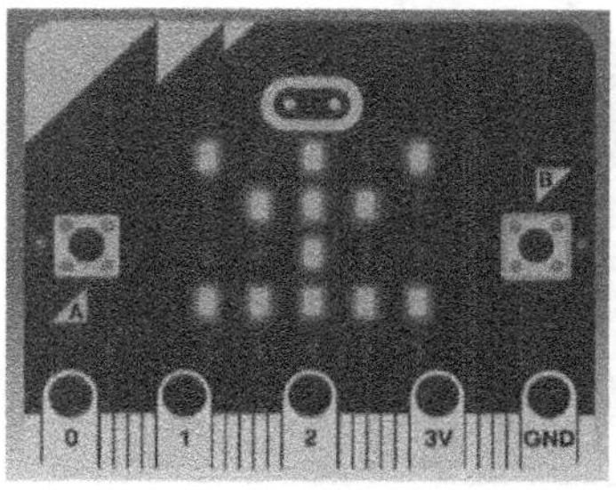  Animation on the micro:bit | 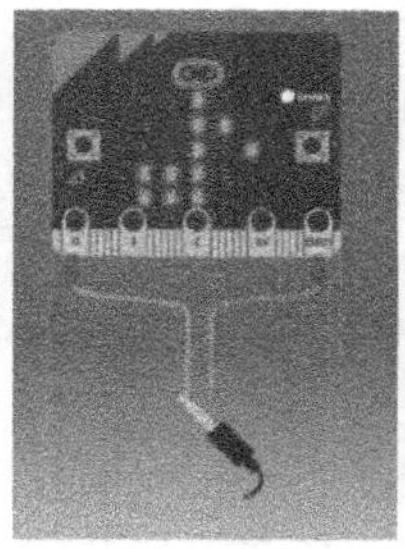 Playing music on the micro:bit | 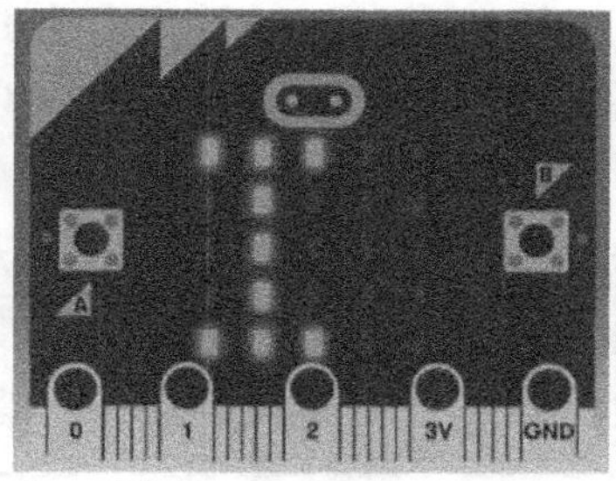  Scrolling message on the micro:bit |

# Chapter 5 – Create with code 2

## Project: Create a scene with layers to show a view from a window

| Name | |
|---|---|

| | Working towards (1) | Meets expected (2) | Exceeding expected (3) | Score |
|---|---|---|---|---|
| **Design a micro:bit project** | • Created a design that partially meets the requirements | • Created a design for a project with one input and one output and explained the algorithm in words | • Created a design for a project with at least one input and one output and provided a detailed explanation of the algorithm in words | |
| **Build and debug a micro:bit project** | • Incomplete or incorrect implementation of the design | • Complete and correct implementation of the design with evidence of debugging | • Complete and correct implementation of a more detailed design with evidence of debugging | |
| **Explain the benefits of collaboration** | • Partial explanation of the benefits of working with a partner | • Explanation of a benefit of working with a partner with an example | • Clear explanations of more than one benefit of working with a partner with examples | |

| Teacher feedback | |
|---|---|
| **Learner response** | |

| Chapter 5: | Lesson 1 Dance algorithm | | |
|---|---|---|---|
| **Learning Objectives** | • Describe logical thinking and its purpose<br>• Describe how tasks can be divided into smaller sections to make them easier to follow and modify | | |
| **New key terms** | **logical thinking** – stepping through a problem to find a solution; logical thinking helps us to make decisions | | |
| **You will need** | • **Optional:** A relevant video of a dance routine to show to the class | **Key pedagogy** | Unplugged |

| Activity | Instructions | Guidance |
|---|---|---|
| **1** | • Introduce logical thinking (Page 54).<br>• Ask learners to complete (Page 46) Task A, 'Is this logical thinking?'.<br>  • Learners choose whether various scenarios are examples of logical thinking. | • Make connections to common tasks at school that require logical thinking, such as sorting toys or planning a piece of writing. |
| **2** | • Introduce the material on dance routines (Page 54).<br>• Optionally, show the class a video of a dance routine.<br>• Use Discuss 1 (Page 54).<br>• Explain that learners are going to design and perform a dance routine. | • Make the connection to culturally relevant dances that will be well known to your learners.<br>• Remember to link the dance moves back to the steps in an algorithm.<br>• You may have dancers in your class. If they are willing, you could discuss their experience and whether they have to learn dance routines. |
| **3** | • Introduce the material on breaking down a task into smaller parts (Page 55).<br>• Put learners into groups and ask them to plan and practise their dance routines. They will complete this planning using (Pages 46 and 47) Task B, 'Plan your dance algorithm'.<br>• Tell learners how much time they have and encourage them to plan accordingly. | • If some learners are reluctant to dance, they can give instructions and make sure the dancers are following the algorithm correctly.<br>• Alternatives to a dance are a rhythm clapping game or a complex handshake.<br>• If some groups finish early, you can direct them to move on to editing their dance. |
| **4** | • Use Discuss 2 (Page 55).<br>• Ask learners to complete (Page 47) Task C, 'Edit your dance algorithm'.<br>  • Learners edit their dance algorithm. | • If there is time, learners could give feedback to other groups to support their changes. |

| Activity | Instructions | Guidance |
|---|---|---|
| 5 | • Use Discuss 3 (📖 Page 55). | • Learners should find that it's easier to work on one section of their routine at a time. They may find it helpful to have separate sections when they want to make changes and need to find the steps to change.<br>• Learners should find that helpful names make it easier to talk about sections. |

**Build on this:**
Join the dances together by putting the groups in order to perform a big dance with the whole class. How many steps does the whole dance have? It would be very difficult to create and understand without breaking it into sections!

| Chapter 5: | Lesson 2 Inputs and outputs with the micro:bit | | |
|---|---|---|---|
| **Learning Objectives** | • Identify the inputs of algorithms<br>• Develop a sequence to produce an output based on an input | | |
| **You will need** | • micro:bit devices, one per pair ideally (if you do not have access to devices, you can use the online simulator at (Website C5L2R1)<br>• Computers to run the MakeCode editor<br>• Make sure you know how to use the MakeCode editor and download code to the micro:bit – see (Website C5L2R2) | **Key pedagogy** | Investigate |

<table>
<tr><th>Activity</th><th>Instructions</th><th>Guidance</th></tr>
<tr><td>1</td><td><ul><li>Remind learners that they used the micro:bit as a night light in Chapter 4, ‘How computers work’.</li><li>Show a physical micro:bit if you have one available.</li><li>Cover the material on inputs and outputs, and inputs and outputs on the micro:bit (Page 56).</li><li>Ask learners to complete (Page 48) Task A, ‘micro:bit inputs and outputs’.<ul><li>Learners identify the inputs on the micro:bit.</li></ul></li></ul></td><td><ul><li>Point out that the LED grid also acts as the light sensor.</li><li>Point out the ‘A’ and ‘B’ buttons.</li><li>Check which version of the micro:bit you have. The images in this book series show the v2 model, which has sound functionality.</li></ul></td></tr>
<tr><td>2</td><td><ul><li>Introduce using inputs and outputs in programs (Page 57).</li><li>Demonstrate creating the example program in MakeCode. Show dragging and dropping blocks.</li><li>Download the code to a micro:bit, if you have one. If not, demonstrate in the simulator.</li><li>Shake the micro:bit and show that the text ‘Hello’ scrolls across it. Explain that the micro:bit only has a small screen so it can’t show all the letters at the same time.</li></ul></td><td><ul><li>Make sure you know how to use the MakeCode editor and download code to the micro:bit – see (Website C5L2R3).</li><li>Highlight MakeCode’s similarity to Scratch. Point out the colour coding of blocks to help learners find them.</li><li>When you write code that uses ‘shake’ the simulator adds a labelled dot that you can click to simulate a shake.</li></ul></td></tr>
<tr><td>3</td><td><ul><li>Ask learners to complete (Pages 48 and 49) Task B, 'Program inputs and outputs'.<ul><li>Learners identify the input and output of example programs in MakeCode and predict what will happen when the code runs. Complete the first example as a class.</li></ul></li><li>Ask learners to test their predictions by running the code in their Workbooks.</li></ul></td><td><ul><li>Learners may need assistance downloading code to the micro:bit for the first time. They will have plenty of opportunities to practise this skill.</li><li>Use the simulator if you do not have micro:bit devices.</li><li>Draw attention to the relationship between the physical buttons on the micro:bit and the code blocks in MakeCode.</li></ul></td></tr>
<tr><td>4</td><td><ul><li>Cover the material on useful micro:bit blocks (Page 57).</li><li>Relate the blocks to the examples in (Pages 48 and 49) Task B.</li></ul></td><td><ul><li>Note that the micro:bit uses milliseconds (ms). 1000 ms is equal to one second.</li></ul></td></tr>
</table>

| | | |
|---|---|---|
| **5** | • Optionally, ask learners who have finished Task B (Pages 48 and 49) to complete (Pages 49 and 50) Task C, 'More program inputs and outputs', which offers more examples like those in Task B. | • This task can be used for learners who finish quickly.<br>• This task provides additional examples that can be used when learners are designing their own projects. |
| **6** | • Use Discuss 4 (Page 57). | • Use this as an opportunity to address any confusion. |

**Build on this:**

Give learners time to make modifications to the example programs.

| **Chapter 5:** | **Lesson 3 Testing for the correct output** | | |
|---|---|---|---|
| **Learning Objectives** | • Describe how to test and debug programs so that they produce the desired output<br>• Explain how programmers use their mistakes to inform the programs that they create | | |
| **New key terms** | **icon** – a small picture that represents a computer program, a computer file, a folder or an action for a program to do | | |
| **You will need** | • micro:bit devices (if you do not have access to devices, you can use the online simulator at (Website C5L3R1)<br>• Computers to run the MakeCode editor | **Key pedagogy** | Investigate |

| Activity | Instructions | Guidance |
|---|---|---|
| 1 | • Use Discuss 5 (Page 58) and recap knowledge. | • Learners should identify these benefits:<br>• It's easier to identify and fix errors when you have just written the code and remember what the code does.<br>• If you know the code was working before, you know that any new problems have been introduced by your latest code. This helps you know where to look to fix an error. |
| 2 | • Introduce the material on common micro:bit mistakes (Page 58).<br>• Ask learners to complete (Page 51) Task A, 'Match the corde to the error'.<br>• Learners match the code to the error.<br>• Share the correct answers with the class and address any confusion. | • You could demonstrate the errors to the class to help with their understanding.<br>• Learners will have repeated opportunities to revisit these errors during the chapter. They may not fully understand until they experience the error themselves. |
| 3 | • Use Discuss 6 (Page 59).<br>• Remind learners how to open the MakeCode editor and tell them to add the 'count up' program (Page 59).<br>• Remind learners to test regularly in the simulator.<br>• Use Discuss 7 (Page 59). | • Explain that the micro:bit adds a small pause between 'show number' blocks so you will see the numbers change.<br>• Some learners may not notice that they need to press the 'A' button to run the code. This is an opportunity to remind them about inputs. |
| 4 | • After testing, tell learners to change their code so that it counts down from 3 to 1 (Page 59). | • They will need to drag the blocks and reorder them in the MakeCode editor. |

| Activity | Instructions | Guidance |
| --- | --- | --- |
| **5** | • Once learners have the countdown code working, ask them to complete (Page 52) Task B, 'Design your icon'.<br>  • Learners design an icon to display on the micro:bit at the end of the countdown.<br>• When they have designed their icon they should update their code and test it.<br>• If learners have MakeCode accounts, their projects will be saved to their account automatically. Otherwise, see the micro:bit section of this Teacher's Guide for saving options. | • Learners could add a 'pause' and a 'clear screen' block at the end of their code to make the icon disappear after a delay. Prompt early finishers to do this by asking: 'How would you make the icon disappear? How long do you want it to show for?'<br>• Learners will not need this project after the lesson but they may want to keep it. |
| **6** | • Ask learners to complete (Page 52) Task C, 'Reflect', where they reflect on their progress.<br>• Use Discuss 8 (Page 59). | • Learners may have mistyped a number or dragged blocks in the wrong order. They may have accidentally deleted blocks – you can mention Ctrl-Z to undo.<br>• Keep a list of common mistakes in a document or on poster paper to use in future lessons. |

**Build on this:**

Learners could create a longer countdown or add an animation with multiple 'show leds' blocks at the end of the countdown.

| Chapter 5: | Lesson 4 Design a micro:bit project | | |
| --- | --- | --- | --- |
| **Learning Objective** | • Create a program that produces an output from an input device | | |
| **You will need** | • Be ready to demonstrate the example projects, either on physical micro:bits or in the simulator.<br>• Example projects:<br>  • Scrolling message project (Website C5L4R1)<br>  • Animation project (Website C5L4R2)<br>  • Playing music project (Website C5L4R3) | **Key pedagogy** | Collaboration |

| | | | |
|---|---|---|---|
| | • **Optional:** Example code for a light sensor (Website C5L4R4) | | |

| Activity | Instructions | Guidance |
|---|---|---|
| 1 | • Re-introduce the project brief and example projects ( Page 60).<br>• Demonstrate the example projects.<br>• Tell students whether they can use music or not. (There is a speaker built in to the micro:bit v2. If you have v1, you could attach speakers, use the simulator or not use sound.) | • Example projects:<br>  • Scrolling message project (Website C5L4R5)<br>  • Animation project (Website C5L4R6) (<br>  • Playing music project (Website C5L4R7) |
| 2 | • Use Discuss 9 ( Page 60).<br>• Ask learners to complete ( Page 53) Task A, 'micro:bit inputs and outputs', based on the discussion and any constraints you put in place.<br>  • Learners list the inputs and outputs they can choose from in their project. | • Discuss whether learners will be able to use a speaker or are limited to the screen.<br>• The 'A' and 'B' buttons are beginner-friendly inputs to use. For micro:bit v2 you also have 'on logo pressed'.<br>• The 'on shake' and 'on loud sound' (microphone, v2) can also be used, as well as 'tilt left/right' and so on (change the 'shake' drop-down).<br>• It is also possible to use the light sensor (example code is provided at (Website C5L4R8) if learners wish to do this because of the night light example in the previous chapter. |

| Activity | Instructions | Guidance |
|---|---|---|
| **3** | • Put learners into pairs and explain that they will be collaborating on a shared project.<br>• Remind learners of the example projects and make sure they understand the kinds of project that are possible.<br>• Discuss the example algorithm in (Pages 53 and 54) Task B, 'Our micro:bit project'.<br>  • Ask them to discuss ideas for their project and then plan it. They should log these ideas in Task B (Pages 53 and 54).<br>• Ask learners to complete (Page 54) Task C, 'My project's input and output'.<br>  • Learners record the input and output they will be using when they have made a decision. | • Observer learners and remind them of the example projects.<br>• You could load the examples onto physical micro:bits and have them available for learners to interact with to remind them what is possible. |
| **4** | • Use Discuss 10 (Page 60).<br>• Tell learners that they will be making their projects in the next lesson.<br>• Ask 'How are you finding working with a partner on this project? Does it help to have someone to share ideas with?' (Page 60). | • Although the micro:bit is a small device, there are lots of things it can do.<br>• Your class may have come up with a variety of ideas. |

**Build on this:**

Learners could discuss their idea with another pair and get feedback. They could then update their designs based on this feedback.

| Chapter 5: | Lesson 5 Program your micro:bit project | | |
|---|---|---|---|
| **Learning Objective** | • Develop a program for a physical computing device that produces an output | | |
| **You will need** | • micro:bit devices (if you do not have access to devices, you can use the online simulator at (Website C5L5R1)<br>• Computers to run the MakeCode editor<br>• **Optional:** Headphones attached to micro:bit v1, if learners need sound | **Key pedagogy** | Collaboration |

| Activity | Instructions | Guidance |
|---|---|---|
| 1 | • Explain to learners that they will be programming their projects this week.<br>• Ask them to look at their Workbooks and remind themselves, in pairs, of their project plans. | • If any learners did not successfully plan a project, they can choose to adapt one of the example projects in the Student's Book. |
| 2 | • Discuss the programs for the example projects (Page 61).<br>• Ask learners to complete (Page 55) Task A, 'Input and output blocks'.<br>  • Learners note down which input and output blocks they will use in their program. | • They should think about the input block that they will use to start their program. They should also think about the output blocks that will be needed to create their planned output (e.g. the 'Show string' block to show a message). |
| 3 | • Remind learners how to open the MakeCode editor and download code to the micro:bit.<br>• Ask learners to complete (Page 55) Task B, 'Testing'.<br>  • Learners write the code for their project, test it regularly, fix bugs and record their progress. | • Refer to the common micro:bit mistakes from Lesson 2.<br>• Learners may want to add spaces to the beginning or end of 'show text' strings or add 'pause' or 'reset' blocks to get the effect they want.<br>• Encourage learners to keep working on their project until it is exactly as they want it. |
| 4 | • Remind learners to save their projects so that they will be able to access them in the next lesson. | • If you are not using MakeCode accounts for students, check the MakeCode guidance at the beginning of this Teacher's Guide for options for saving projects. |
| 5 | • As learners finish their projects, ask them to complete (Page 56) Task C, 'Instructions', and Task D, 'Advice'.<br>  • Learners write instructions and advice for others interacting with their project. | • Some learners may not have time to complete these tasks. |
| 6 | • Use Discuss 11 (Page 62). | • Relate the discussion to the importance of testing.<br>• Mention the importance of collaboration and working together to solve problems. |

| Activity | Instructions | Guidance |
|---|---|---|
| **Build on this:**<br>If learners have more time, they can add scripts to their program to handle additional inputs. They can make longer sequences, such as longer scrolling banners, animations or music. They can also try using more than one output, such as sound and scrolling text. | | |

| Chapter 5: | Lesson 6 Showcase your micro:bit project | | |
|---|---|---|---|
| **Learning Objective** | • List the benefits of collaboration in programming | | |
| **You will need** | • Learners' final projects<br>• micro:bits/simulators | **Key pedagogy** | Showcase |

| Activity | Instructions | Guidance |
|---|---|---|
| **1** | • Give pairs time to download their project onto a micro:bit and complete (Page 57) Task A, 'Benefits of collaboration', where they list the benefits of collaboration. | • Remind learners that they should refer to Task A during their showcase. |
| **2** | • Run the showcase, asking pairs to demonstrate their projects. | • Prompt pairs using (Page 63), if necessary. |
| **3** | • Ask learners to complete (Page 57) Task B, Celebration icon', and (Page 58) Task C, 'Reflection'.<br>  • Learners create a paper-based celebration icon and reflect on the success of their project. | • If you have time, choose one or more of the celebration icons that learners design and create a micro:bit project to display them.<br>• You can use full-screen mode in the micro:bit simulator to show a large version. Use this if you are displaying a student's work on the projector. |

| Activity | Instructions | Guidance |
| --- | --- | --- |
| **4** | • Celebrate learners' successes. They have learned how to create a micro:bit project that produces an output based on an input. They have worked together and showcased their projects to the class.<br>• Ask learners to reflect to wrap up the chapter by asking: 'How did it feel to work with a partner on this project? Were you able to stick to your plans or did you change them?' (Page 64). | |

**Build on this:**

Run a showcase with another class or with parents and guardians to celebrate learners' hard work during this chapter.

# Chapter 6 – Connect the world

## Project: Create a visualisation of your school network

## Chapter overview

In this chapter, learners will recall network hardware that was covered in Stage 1. They will then think about familiar services that are available through networks and the internet, and discover different types of file that they can access and store. Next, they will have a class debate on whether networks are good or bad. Finally, learners will tour the school to spot networked hardware and create a visualisation of the school network, using a medium of their choosing.

| Chapter 6: Lesson summary | | |
|---|---|---|
| **Lesson** | **Learning Objectives** | **Summary** |
| **1** | • List different types of file that can be stored on a computer's hard drive | Learners explore different file types that exist in computing. They investigate which ones they have used. |
| **2** | • Identify common services available on a network, including file sharing and access to the World Wide Web | Learners investigate services provided by the internet and discuss screen time. They also think about how a network can be used to share resources. |
| **3** | • List the advantages and disadvantages of a network | Learners explore the benefits and dangers of networks. This culminates in a class debate on whether networks are good or bad. |
| **4** | • Identify the networked hardware used in school | Learners discover some of the physical parts that make up a network and go to find them around the school. While touring the school, they sketch out a 'node and edge' diagram of the network to be used later. |
| **5** | • Create a visualisation of the school network | Learners build a model – a visualisation of the school network using materials chosen by them or their teacher. |
| **6** | • Showcase a visualisation of the school network | Learners have the opportunity to showcase their visualisations of the school network to the class. They can observe the potential similarities in structure but the uniqueness of each person's creativity. |

| End of chapter project | Create a visualisation of your school network |
|---|---|
| **Example ideas that learners could design:** | 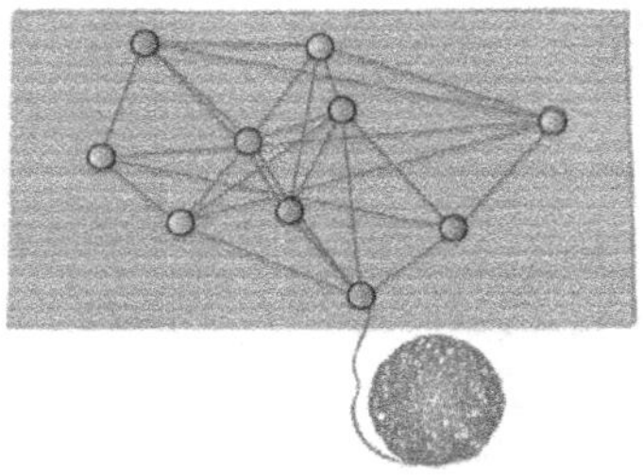 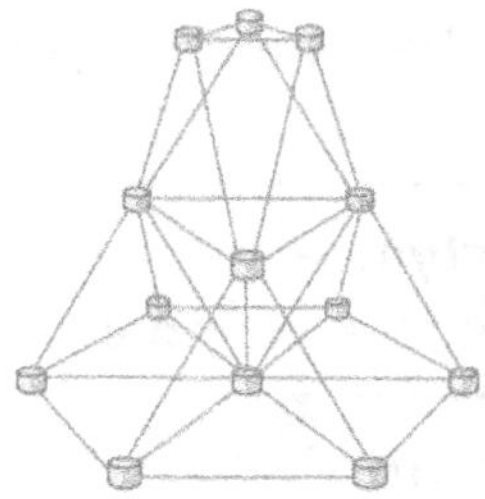  |

## Chapter 6 – Connect the world

### Project: Design a webpage on a topic of your choice

| **Name** | |
|---|---|

| | **Working towards (1)** | **Meets expected (2)** | **Exceeding expected (3)** | **Score** |
|---|---|---|---|---|
| **Draw the school network** | • The school network was drawn with partial accuracy | • The school network drawing was mostly accurate | • The school network drawing was accurate and detailed | |
| **Refine your drawing** | • The drawing has been repeated, but not refined | • The drawing demonstrates refinement (it is neater and improved) | • The drawing has been refined and improved, with a high level of accuracy and detail | |
| **Build your visualisation** | • A partial visualisation has been built | • A visualisation of the school network has been built | • A highly detailed and accurate visualisation of the school network has been built | |

| **Teacher feedback** | |
|---|---|
| **Learner response** | |

| Chapter 6: | Lesson 1 File types |
|---|---|
| **Learning Objective** | • List different types of file that can be stored on a computer's hard drive |
| **New key terms** | **file type** – the way a computer stores a file. Every file on a computer has a specific type; for example, 'document', 'video' or 'image' file types |
| **Key pedagogy** | Get creative |

| Activity | Instructions | Guidance |
|---|---|---|
| **1** | • Introduce the project brief that learners will be working towards in this chapter (Page 65).<br>• Learners will be creating a visualisation of the school network. | • You could link back to Stage 2, Chapter 6, where learners painted a visualisation of what a network might look like. |
| **2** | • Explain that computers can be used for different activities (Page 66).<br>• Use Discuss 1 (Page 66). | • There are five different activities taking place with computing devices:<br>• listening to music<br>• taking a photo<br>• recording a video<br>• writing a document<br>• drawing a picture. |
| **3** | • Ask learners to complete (Page 59) Task A, 'What have you used a computer for?'.<br>• Learners draw pictures of all the activities they have used a computer for.<br>• If there is time, learners could share their images with a partner. | • Focus on activities that you have completed in school, such as using a Scratch project, creating a presentation and typing on the keyboard. Learners may have also played games or watched videos if they have a device at home. |
| **4** | • Explain that different activities on a computer have different file types (Page 67). For example, a video is saved as a video file.<br>• One of the reasons for this is so that a computer knows which piece of software to load to display the file correctly. | • The technical knowledge behind this is that each file type has a particular encoding and meta data. This is used to display the information correctly and select the correct software to load. Learners do not need to know this at Stage 3, but it is worth having this information in case you are asked. |

| Activity | Instructions | Guidance |
|---|---|---|
| 5 | • Ask learners to complete (Page 60) Task B, 'Match the file types to the programs'.<br>  • Learners match the type of file to the software that might load to display the contents of the file.<br>• You may wish to ask learners to peer-mark their answers using the answer sheet. | • The slide deck example has been completed for them. They should know what presentation software is at this stage but they may not be familiar with the term 'slide deck'. You can explain that a slide deck is just a collection of slides in a presentation. |
| 6 | • Explain that you can save files on your computer, or through a network, including the internet.<br>• Use Discuss 2 (Page 67). | • If learners have created an account online with Scratch, they can use this as an example of saving files through the internet. If learners have been using online word processing and presentation software in their Content creation lessons, this will be another good example.<br>• The next lesson will cover saving files through a network in more detail. |
| **Build on this:**<br>Take a look together at some of the files on your school network and see which are documents, videos and sound files. | | |

| Chapter 6: | Lesson 2 Common services |
|---|---|
| **Learning Objective** | • Identify common services available on a network, including file sharing and access to the World Wide Web |
| **New key terms** | **streaming service** – online provider of entertainment, such as websites or apps that allow you to watch movies |
| **Key pedagogy** | Classroom talk |

| Activity | Instructions | Guidance |
|---|---|---|
| **1** | • Show learners the illustration that demonstrates some different activities that can take place online using a computing device (Page 68).<br>• Introduce the term 'streaming service' and use Discuss 3 (Page 68).<br>• Point out the 'stay safe' content about there being age restrictions for streaming services. Remind learners to always ask an adult before viewing online content. | • Different households will have different rules for using streaming services. It is important to be cautious about this in your discussions. |
| **2** | • Ask learners to complete (Page 61) Task A, 'Screen time survey'.<br>  • Learners complete a screen time survey. The survey asks how long they think they should be allowed to use a computing device for different activities.<br>• Hold a discussion with the class about the level of screen time they think they should be allowed to have. Use Discuss 4 (Page 68). | • Some learners may want to be allowed as long as they want; some might be happy with 30 minutes.<br>• The music question might be a different discussion because you can do other activities while you listen to music and don't have to stare at a screen. |

| | | |
|---|---|---|
| **3** | • Read the short paragraph about screen time (📖 Page 69). | • When discussing screen time, every household will have their own rules. The latest advice is that we should take regular breaks. We should also monitor the types of activity that we are using a device for. For example, spending eight hours watching YouTube videos is not healthy, but spending eight hours doing different activities for work and entertainment throughout the day is a better use of screen time.<br>• Go to the World Health Organization website, select Newsroom, News releases and search 'Children need to play more' for an article that provides extra information about screen time. You could read this in advance and use the information as prompts in a screen time discussion. |
| **4** | • Introduce how networks can be used to share resources such as printers (📖 Page 69). Use the example diagrams to show how a printer can be shared when it is attached to a network.<br>• Ask learners to complete (📓 Pages 61 and 62) Task B, 'Share resources'.<br>  • Learners find solutions to resourcing problems by creating a network. | • You may decide to do the Workbook task with the whole class as part of a discussion. You could draw various solutions on the whiteboard and ask learners to pick the correct one. Use the answers in the answer sheet to support this.<br>• It is important to note that we aren't asking learners to draw technically accurate network diagrams. They are simply drawing models to help them understand that devices can be connected to share resources. |

| | | |
|---|---|---|
| 5 | • Use Discuss 5 (📖 Page 69). | • Learners may give answers such as they are good when you want to share files with someone else. They could say they are bad because they don't work all the time.<br>• If there have been any recent issues in your school or local area relating to network issues or advantages, you could share it here.<br>• This question builds nicely into the next lesson. |

**Build on this:**

Ask learners to create a chart to show how much time they spend on different activities using computing devices in a typical day. You could also complete one for yourself to show learners.

| **Chapter 6:** | **Lesson 3 Networks – good or bad?** | | |
|---|---|---|---|
| **Learning Objective** | • List the advantages and disadvantages of a network | | |
| **You will need** | • **Optional:** You may wish to find a recent example in the news for your location that relates to a network not being secure and getting hacked. | **Key pedagogy** | Classroom talk |

| Activity | Instructions | Guidance |
|---|---|---|
| 1 | • Start the lesson with a quick recap of the good things that networks and the internet can allow us to do. Use the learnings from the previous two lessons for this.<br>• Use Discuss 6 (📖 Page 70). | • Examples that learners might give include that networks help us to:<br>  • share resources such as printers or the internet<br>  • access streaming services<br>  • create things online such as Scratch projects<br>  • share files.<br>• List learners' answers on the board. You may want to distribute sticky notes to which learners can add their ideas. |

| Activity | Instructions | Guidance |
|---|---|---|
| 2 | • Use the illustrations and examples to list some of the negative things that can happen with a network (Page 70). | • You may have an example of something that has happened recently in the news around network safety. For example, bank details might have been stolen from customers meaning that customers had to cancel their bank cards to keep them safe. |
| 3 | • Ask learners to complete (Page 63) Task A, 'Networks – good or bad?'.<br>  • Learners list two good things and two bad things about networks. They will use this for a class debate in the next activity. | • Learners have had lots of examples to choose from during the previous few lessons. Draw on those if learners need support. |
| 4 | • Hold a class debate on whether networks are good or bad. You can decide to run this, based on your class. One model would be to split the class into pairs. One of each pair is on the 'good' side and the other is on the 'bad' side. Give them time to practise arguing for their side with their partner. Next, move them into larger groups or choose learners to debate in front of the class. | • You know your learners best, so decide on a scenario for this debate based on the needs of your class. Some might not be confident enough to speak in front of the class but be happy to speak to a partner, for example. |
| 5 | • Take a class vote, deciding whether networks are good or bad. | • You could do this with hands up or by saying one side of the room is good and the other is bad. You could give learners the opportunity to scale their point of view so, if they think it is a bit of both, they can stand in the middle of the room. |
| 6 | • Use the text (Page 70) to confirm that networks are neither good nor bad and that we should embrace technology but know the risks.<br>• Use Discuss 7 (Page 70). | • Learners should be comfortable enough to give an answer at this point, but draw on examples from the start of this lesson if needed. |

| Activity | Instructions | Guidance |
|---|---|---|
| **Build on this:**<br>If there is time, learners could do a quick data visualisation activity. Learners stand in a location based on their vote and then you take a picture to visualise their point of view. You could then use the picture as a starting point for a discussion about how visualisation is a useful tool to see results quickly and come to a conclusion. Understanding the advantages of visualisation is useful for both this and the Power of data chapters. | | |

| Chapter 6: | Lesson 4 The network in your school | | |
|---|---|---|---|
| **Learning Objective** | • Identify the networked hardware used in school | | |
| **New key words** | **visualisation** – using creative ways to display information | | |
| **You will need** | • **Optional:** Photographs of key areas in the school that visibly show network hardware, such as the server room if you have one | **Key pedagogy** | Get creative |

| Activity | Instructions | Guidance |
|---|---|---|
| 1 | • Tell learners that there are physical parts that are used to create a network. Show them the photos of different parts of a network (📖 Page 71). | • The images are there as a guide. Your school may not have these exact network components; however, there may be parts that are similar. |
| 2 | • Remind learners of the project brief for this chapter. You can show learners the student exemplars (📖 Page 65).<br>• Ask learners to find (📖✎ Page 64) Task A, 'Draw the school network'. This gives them space to draw their first draft of the school network. They will complete this as they tour the school. | • Use the example drawing in the Workbook as a guide. Tell learners that their drawings do not have to be perfect. Simple dots and lines are fine, with squares to show the different rooms.<br>• Instruct learners to label their drawings as you go on the tour as this will help them when they make their models and do their presentations. |

| Activity | Instructions | Guidance |
|---|---|---|
| **3** | • Tour the school, stopping to look at all the physical parts of the school network along the way. | • If your school has a network technician or a person responsible for the school network, it might be helpful to ask them to be part of the tour. They will be able to spot things that you might miss and give more context for learners.<br>• Give learners plenty of time to draw the parts of the network as they go.<br>• If touring the school is not possible, you could prepare a slideshow in advance with photos from your school. |
| **4** | • Now that you are back in the classroom, ask learners to create a new version of their drawing (Page 65) in Task B, 'Refine your drawing'. This should be a slightly neater version as they are now back at their desks. | • You could provide a slideshow of images of the network parts around the school to remind learners of what they have seen. |
| **5** | • At the end of the lesson, give learners some time to share their work with a partner and see if they have missed anything out of their designs. Use Discuss 8 (Page 71). | • To pick up anything that has been missed, encourage learners to think about the tour step by step, compare their drawings with others and use the information in the Student's Book |

**Build on this:**

Learners could improve their designs based on feedback from a partner.

| Chapter 6: | Lesson 5 Create a visualisation | | |
|---|---|---|---|
| **Learning Objective** | • Create a visualisation of the school network | | |
| **You will need** | • Materials for learners to build their visualisation models. This could be:<br>• sticky tac and straws<br>• string, glue and paper<br>• cotton wool balls and dried spaghetti<br>• painting materials<br>• recycled plastic or boxes<br>• fabric cuttings | **Key pedagogy** | Get creative |

 

| Activity | Instructions | Guidance |
|---|---|---|
| 1 | • Read through 'What is a visualisation?' (📖 Page 72).<br>• Remind learners of the drawings that they created of the school network during their tour in the previous lesson (📖 Page 65). | • If learners missed the previous lesson, you could provide them with an example drawing or they could partner up with someone in the class and use their drawing for their model. |
| 2 | • Give learners time to build their models using the drawings that are in their Workbooks as a guide. | • You may want to create your own version of a model to demonstrate to the class.<br>• For example, if you are using cotton wool balls and dried spaghetti, you might want to give learners some tips about joining them together without breaking the spaghetti. For example, be gentle and twist slowly. |
| 3 | • Remind learners that visualisations show information; they should be easy for someone else to understand.<br>• Allow some time for learners to provide feedback on each other's work. | • It would be interesting to start to notice any similarities or differences between the models. Are they all the same? |
| **Build on this:**<br>Learners could improve their designs based on feedback from a partner. | | |

| Chapter 6: | Lesson 6 Showcase your visualisation | | |
|---|---|---|---|
| **Learning Objective** | • Showcase a visualisation of the school network | | |
| **You will need** | • The visualisation models that learners created in the previous lesson | **Key pedagogy** | Showcase |

| Activity | Instructions | Guidance |
|---|---|---|
| 1 | • Let learners know that they will be showcasing their visualisations in this lesson.<br>• Help them practise how they will showcase their models, reminding them of the criteria for the showcase (Page 73). | • Give learners time to think of a short script of what they would like to say about their models.<br>• Give them time to practise their demonstrations.<br>• Encourage learners to think of how they will position their model and themselves during the presentation so that the class can see the models. |
| 2 | • Showcase the visualisations. | • You could run this activity as a whole class session or in groups.<br>• Encourage learners to pay attention to the inventive ways the different materials were used and to take notes of any visualisations they thought were very easy to understand. |
| 3 | • Ask learners to complete (Page 66) Task A, 'Reflection'.<br>  • Learners reflect on the project. They can share their reflections with a partner.<br>• Celebrate all that learners have achieved in this chapter. As well as investigating how the world is connected, they should also feel more connected to their school and how the school network works. | • If learners took notes during the showcase, you should encourage them to share their thoughts as part of the reflection.<br>• You could take pictures of the models to become part of a display or for learners to stick into their Workbooks as it is unlikely that models created with these materials will be able to be kept for long. |

**Build on this:**

Learners should think about where else might have similar networks to their school. They can list some other places in the local area such as other schools, offices, government buildings or public libraries. What will be similar?

# Chapter 7 – The power of data

## Project: Use data to solve a problem in your school or local community

### Chapter overview

In this chapter, learners will understand the characteristics of good data and how it can be used to solve problems. They will practise using spreadsheets to input, format and visualise data. They will then use their skills to help them solve a problem in their school or local community.

**Chapter 7: Lesson summary**

| Lesson | Learning Objectives | Summary |
|---|---|---|
| 1 | • Identify problems that can be solved through the collection and interpretation of data<br>• Discuss data characteristics that are required to solve problems, including their accuracy, completeness and timeliness | Learners discuss the importance of data to make important decisions. They identify data that is complete, accurate and timely. Learners practise collecting good data via completing a class tally chart. They then reflect on the process they used. |
| 2 | • Define the terms 'cell', 'row' and 'column', and enter data into a spreadsheet<br>• Format cells in a spreadsheet according to their purpose, such as currency | Learners explore spreadsheets and their uses. They enter data into a spreadsheet, then format it into the correct data types. Learners use the formatted data to answer questions. Learners practise entering and formatting data using example data sets. |
| 3 | • Investigate different methods for representing discrete and categorical data | Learners explore definitions of discrete and categorical data, then investigate how to use them in charts. They answer questions for data visualised in pie charts and bar charts. Learners enter data then create their own chart, before analysing and documenting the results. |
| 4 | • Choose a problem that can be solved by collecting discrete and categorical data, using a computing device<br>• Design questions for gathering discrete and categorical data, using a computing device | Learners take part in a class discussion to identify problems that can be solved by data. They look at student examples and decide which problem they should choose for their project. Learners plan their spreadsheet, detailing the formatting required. They design a form to collect their data and carry out their data collection. |
| 5 | • Collect and input data<br>• Visualise data using a chart | Learners input their data into a spreadsheet and format their data. They look at charts used in student examples, then insert a chart into their data. |

| | | |
|---|---|---|
| **6** | • Showcase a visualisation that solves a problem | Learners practise their presentations, using the tips provided. They take part in a showcase session to present their data and answer questions. |

| End of chapter project | **Use data to solve a problem in your school or local community** |
|---|---|
| **Example ideas that learners could design:** | 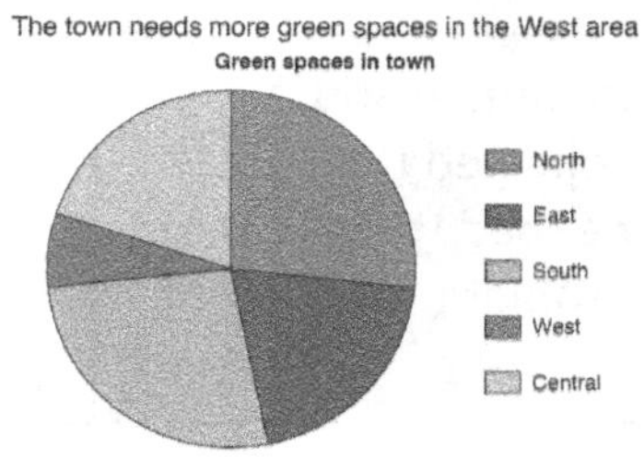  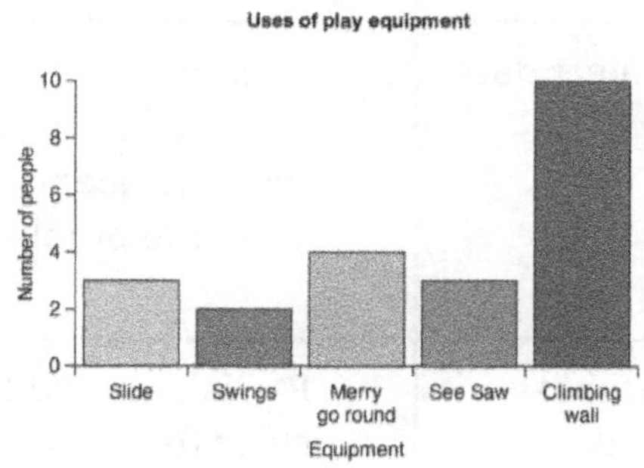  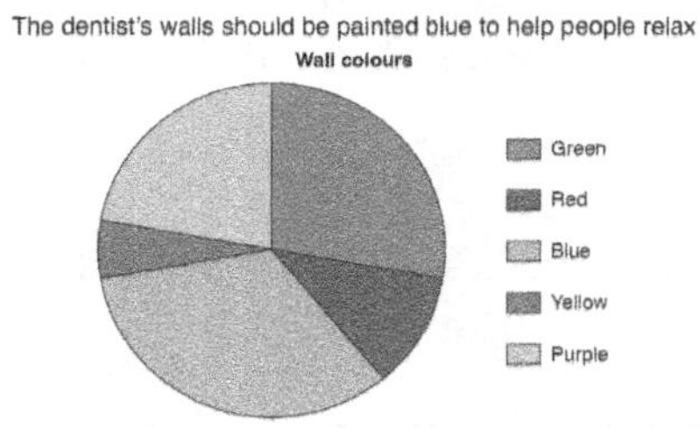  |

# Chapter 7 – The power of data

## Project: Plan a dream class celebration

| Name | |
|---|---|

| | Working towards (1) | Meets expected (2) | Exceeding expected (3) | Score |
|---|---|---|---|---|
| **Collect data** | • Identified and collected some of the data needed to solve the problem | • Identified and collected the data needed to solve the problem | • Identified and collected the data needed to solve the problem. Discussed the plan with others | |
| **Visualise data** | • Partially created a chart to visualise the data | • Created a chart to visualise the data | • Created a chart to visualise the data. Experimented with different types of chart to choose the most appropriate solution | |
| **Present data** | • Showcased the results with limited explanation of the visualisation | • Showcased results using the visualisation as a prop to explain the data driven decision | • Showcased results using the visualisation as a prop to explain the data driven decision. Shared insights with those impacted by the decision | |

| | |
|---|---|
| **Teacher feedback** | |
| **Learner response** | |

| Chapter 7: | Lesson 1 Good data |
|---|---|
| **Learning Objectives** | • Identify problems that can be solved through the collection and interpretation of data<br>• Discuss data characteristics that are required to solve problems, including their accuracy, completeness and timeliness |
| **Key pedagogy** | Classroom talk |

| Activity | Instructions | Guidance |
|---|---|---|
| **1** | • Introduce learners to the end of chapter project (Page 75). Remind learners that they already know about data and using data to make decisions from 'The power of data' chapters in Stages 1 and 2.<br>• Talk through the market stall illustration (Page 76).<br>• Use Discuss 1 (Page 76). | • For the discussion question, learners might answer:<br>  • so that they can work out how much money they have made<br>  • so that they know when to reorder more fruit<br>  • so that they can see what is popular or unpopular. |
| **2** | • Explain that data can be used to make very important decisions.<br>• Use the illustrations to explain how good data needs to be complete, accurate and timely (Page 77).<br>• Use Discuss 2 (Page 77). | • You could mention that timeliness of data includes being available at the point you need it, which includes being up to date.<br>• Learners may have heard you talking about the importance of the class register and you can prompt them to recall this.<br>• Use the illustrations as a guide to help learners answer. Potential answers could be:<br>  • If there was a fire alarm, you wouldn't know who was at school or not.<br>  • If you were planning a lesson, you wouldn't know how many resources to make.<br>  • If you needed to report absences or give out attendance awards, you wouldn't know who to include.<br>  • The school meal preparation staff wouldn't know how many meals to make. |

| Activity | Instructions | Guidance |
|---|---|---|
| **3** | • Ask learners to complete (Page 67) Task A, 'Identifying good data'.<br>  • Learners identify why the data in the examples is not good data. | • If learners need a little help, you could point out that:<br>  • it is the end of the week, but there is no data for Thursday or Friday<br>  • the children are remembering things differently<br>  • there are 30 students in the class, yet only 7 have answered the question. |
| **4** | • Remind learners that they have completed tally charts before when they planned a dream class celebration in Stage 1. Remind learners that tally charts are a visual way to show data, (Page 77).<br>• Ask learners to complete (Page 68) Task B, 'Collecting good data'.<br>  • Learners collect and record data from their classmates. | • Learners can use any method they choose to collect the data, but they must get good data so their data must be complete. |
| **5** | • Ask 'Was it easy to capture good data in the Workbook activity? What methods did you use to make sure your data was complete?' (Page 77). | • Learners might have found it hard to collect good data if they all went around the classroom asking each other in turn. They might have forgotten who they had asked (or even forgotten to include themselves).<br>• If learners took a more structured approach, they would have found it easier. They could have done it one table at a time or done a show of hands as a class. |

**Build on this:**
Learners could make a poster to be displayed in classrooms to stress the importance of complete, accurate, timely school register data for the school.

| Chapter 7: | Lesson 2 Spreadsheets | | |
|---|---|---|---|
| **Learning Objectives** | • Define the terms 'cell', 'row' and 'column', and enter data into a spreadsheet<br>• Format cells in a spreadsheet according to their purpose, such as currency | | |
| **New key terms** | **spreadsheet** – software used to organise and analyse data<br>**row** – in a spreadsheet, a line of cells from left to right, identified with a number<br>**column** – in a spreadsheet, a line of cells from top to bottom, identified with a letter<br>**cell** – individual box in a spreadsheet for storing one piece of data | | |
| **You will need** | • Computers with access to spreadsheet software. Google Sheets has been used in the examples. | **Key pedagogy** | Develop practical skills |

| Activity | Instructions | Guidance |
|---|---|---|
| **1** | • Remind learners what they already know about other software and their similarities with screen layout and shortcuts. | • Learners have used a word processor (likely Google Docs) and presentation software (likely Google Sheets) in Stage 3, Chapter 2, 'Content creation'. Learners have used data capture forms to collect data in Stage 1. They also looked at tally charts as a method to capture data and answer questions. |
| **2** | • Introduce spreadsheet software (📖 Page 78). Explain that spreadsheets are often part of a group of apps that include word processors and presentation software, so the layout and shortcuts can be very similar.<br>• Use the screenshots to show a row, a column and an individual cell. | • You could open spreadsheet software (such as Google Sheets) and highlight each element as you talk through the images. |

| | | |
|---|---|---|
| **3** | • Demonstrate how to enter the data table from Build 1 (Page 78).<br>• Set up learners on the computers with a new spreadsheet open. Learners should now enter the same data from the Student's Book.<br>• Ask learners to complete (Page 69) Task A, 'Answering questions'.<br>  • Learners answer questions about the data they entered from the Student's Book. | • Mention that you are starting in column A and row 1. The formatting instructions in the next step refer to the column letters.<br>• Enter the data as shown in the Student's Book. Learners will format the data later.<br>• Remind learners to use their typing skills when entering the data. In Stage 3, Chapter 2, 'Content creation', they learned to position their fingers on the home keys using the bumps on the keys. |
| **4** | • Introduce data formatting and the types of formatting required for each of their data columns.<br>• Demonstrate formatting the data using the data table on (Page 79).<br>• Follow the steps in the Student's Book, making sure to use the same language, for example: 'Now I'm clicking on A to highlight column A'. | • Use your local default currency for the 'coins collected' column.<br>• You can use a different date format if more appropriate for the 'date complete' column such as mm/dd/yyyy. It is likely that your localisation settings will default to the correct format for your locale. |
| **5** | • Ask learners to complete Build 2 (Page 79).<br>• Ask learners to complete (Page 70) Task B, 'Formatting data'.<br>  • Learners enter the data in the tables in the format shown. | • Learners should practise entering the data in the Workbook into a spreadsheet and then formatting it to match.<br>• Learners might need reminding to highlight a column before applying the formatting. |

**Build on this:**

Learners could come up with their own datasets that include columns to be formatted. They should explore the format options when designing their datasets, then swap with someone else to input and format the data.

 

| Chapter 7: | Lesson 3 Data charts | | |
|---|---|---|---|
| **Learning Objective** | • Investigate different methods for representing discrete and categorical data | | |
| **New key terms** | **categorical data** – information that can be divided into categories, for example, types of drink<br>**discrete data** – information that can only have certain values, usually whole numbers, for example, counts of animals | | |
| **You will need** | • Computers with access to spreadsheet software | **Key pedagogy** | Develop practical skills |

| Activity | Instructions | Guidance |
|---|---|---|
| **1** | • Define categorical data and list the examples (📖 Page 80).<br>• Use Discuss 3 (📖 Page 80). | • Guide learners to think of examples relevant to them: perhaps hair or eye colour, pet types, nationality, or favourite foods, colours or hobbies. |
| **2** | • Define discrete data and list the examples (📖 Page 80).<br>• Use Discuss 4 (📖 Page 80). | • Examples that can be collected in the classroom could be shoe sizes, number of languages spoken, count of books on a shelf, or number of siblings or pets. |
| **3** | • Explain why visualising data is useful and introduce pie charts (📖 Pages 80 and 81).<br>• Ask the question: 'Which continent has the most countries?' (📖 Page 81).<br>• Introduce bar and column charts (📖 Page 81).<br>• Ask the question: 'How many countries are in Asia?' (📖 Page 81).<br>• Ask learners to complete (📝 Page 71) Task A, 'Chart visualisation'.<br>  • Learners interpret bar graphs and pie charts. | • Remind learners that they have used visualisations before, when they created a visualisation of their school network in Chapter 6.<br>• You could ask learners more questions about the example pie and bar charts, if needed.<br>• Learners could mark the answers to the Workbook in pairs, using the charts to explain how they arrived at their answers. |

| Activity | Instructions | Guidance |
|---|---|---|
| 4 | • Demonstrate how to create the chart using the Build 3 (📖 Page 81).<br>• Ask learners to complete Build 3 (📖 Page 81).<br>• Ask learners to complete (📓 Page 72) Task B, 'Reading my chart'.<br>  • Learners answer questions about the chart they created in the build tasks.<br>• Discuss the results as a class. | • The instructions have been written for Google Sheets so, if you are using different spreadsheet software, you might need to alter them slightly and highlight the changes during your demonstration.<br>• The default chart type might be different for each learner (based on their most recent usage). Some may have a column chart already shown, but others will need to choose it from the chart type settings.<br>• The task shows that creating a chart makes it easier to answer questions about the data. |
| 5 | • Ask 'Why are charts a good way of presenting data?' (📖 Page 81). | • You could remind learners that their answers are in their Student's Books. They should be looking for things like:<br>  • charts can display different types of data<br>  • they are visual and can be quickly read<br>  • patterns can be seen clearly. |

**Build on this:**

Learners can change the chart type of their build task to a pie chart and bar chart to get used to editing chart settings. They should reflect on how easy it is to answer the questions with each chart type.

| Chapter 7: | Lesson 4 Collect your data |
|---|---|
| **Learning Objectives** | • Choose a problem that can be solved by collecting discrete and categorical data, using a computing device<br>• Design questions for gathering discrete and categorical data, using a computing device |
| **Key pedagogy** | Classroom talk |

| Activity | Instructions | Guidance |
| --- | --- | --- |
| 1 | • Re-introduce the project brief and talk through the example projects and what they show (Page 82).<br>• Use Discuss 5 (Page 82).<br>• Ask learners to complete (Page 73) Task A, 'Problem discussion'.<br>  • Learners write down problems in their school or community, as raised in the discussion. | • Learners should use the notes on the board to complete the Workbook activity.<br>• If there are a lot of problems discussed and learners are not able to write all of them, they could choose three or four that they think are most interesting.<br>• When discussing, encourage learners to think about whether data could help solve the problem by either confirming it is an issue or identifying a solution.<br>• Set any boundaries for data collection; for example, whether the audience/location is accessible during school time. |
| 2 | • Use Discuss 6 (Page 82). | • You could identify groups of learners to work together, based on their answers.<br>• Learners might have strong reasons why they want to choose a particular problem. Some learners may want to solve the problem they originally suggested, but they can choose a different problem too. |

| Activity | Instructions | Guidance |
|---|---|---|
| **3** | • Discuss the student examples (Page 83). Identify the problem, data required, format of the data and how the student collected the data.<br>• Ask learners to complete (Page 74) Task B, 'Spreadsheet design'.<br>  • Learners write down their plan for collecting their data.<br>• Use Discuss 7 (Page 83). | • You could ask learners whether they think the data captured in the student examples is useful to help solve the problem.<br>• Observe learners as they create their plans. They may need to revisit the Student's Book tasks from earlier in the chapter to remind themselves of types of field and formatting options.<br>• Remind students of any limits you have set around who they can ask questions to and what observations they can make.<br>• Not all learners have to share their plan, but it would be good to get a few examples of plans that are different from each other. |
| **4** | • Ask learners to complete (Page 75) Task C, 'Collect your data'.<br>  • Learners collect and record their data. | • Allow plenty of time for learners to gather their data in their Workbooks.<br>• If the plan involves collecting data outside of lesson time, you could set this as homework. |

**Build on this:**

You could plan a class outing to collect data from the local area or arrange a class visit by a group of people who could provide data on a specific topic.

| Chapter 7: | Lesson 5 Visualise your data | | |
|---|---|---|---|
| **Learning Objectives** | • Collect and input data<br>• Visualise data using a chart | | |
| **You will need** | • Computers with access to spreadsheet software<br>• **Optional:** Printer to print out charts and glue sticks to stick them in Workbooks | **Key pedagogy** | Develop practical skills |

| Activity | Instructions | Guidance |
| --- | --- | --- |
| **1** | • Ask learners to remind themselves of their plan and the data they collected.<br>• Get learners set up at computers with spreadsheet software.<br>• Ask learners to complete Build 4 (Page 84) to enter and format their data from Workbook Task C (Page 75). | • You could get learners to discuss their plan and data collected with someone else. This will help remind them of their work and give them practice for the presentation in the next lesson.<br>• Remind learners to use their Student's Book if they need a reminder of how to format columns. |
| **2** | • Talk through the student examples of chart visualisations and what their analysis told them (Page 84).<br>• Ask learners to complete Build 5 (Page 84). | • The student examples build on those from the previous lesson.<br>• Encourage learners to experiment with the different types of chart they have learned about to find the one that best solves the problem.<br>• Remind learners they need to highlight their data before inserting a chart.<br>• Remind learners to save their spreadsheets as they will need to present them in the next lesson. |
| **3** | • Ask learners to complete (Page 76) Task A, 'My chart'.<br>    • Learners describe their chart choices.<br>• Optionally, print out charts for learners to stick in their Workbooks. | • Tell learners that the answers to this activity will be useful when presenting their project in the next lesson. |

**Build on this:**

Learners could use the chart to analyse their data further. Does the data suggest a potential solution that they didn't expect (for example, Bisma in the student example found that Mondays were a great day to target as Monday was the day more people travelled by car).

| Chapter 7: | Lesson 6 Showcase your results | | |
|---|---|---|---|
| **Learning Objective** | • Showcase a visualisation that solves a problem | | |
| **You will need** | • The spreadsheets from the previous lesson<br>• There may be people from the local community that you would like to invite to the showcase. | **Key pedagogy** | Develop practical skills |

| Activity | Instructions | Guidance |
|---|---|---|
| **1** | • Tell learners the first part of the lesson is time for them to practise their presentations.<br>• Talk through the tips for presenting (Page 85). | • Learners can plan by reading, making notes or practising in pairs.<br>• You could demonstrate how to drag a chart to make it bigger or how to position it with the data captured still visible on the screen. |
| **2** | • Conduct the showcase. | • If you have invited people from the local community, you could ask them to briefly introduce themselves.<br>• Ask learners a couple of questions from their 'spreadsheet design' and 'my chart' activities.<br>• If you have invited community members, ask them to give brief feedback to learners.<br>• If you have not invited community members, give each class member another person to provide brief, kind feedback to. |
| **3** | • Ask learners to complete (Page 77) Task A, 'Reflection', where learners reflect on the project.<br>• Discuss the responses to question 3 (Page 77). | • You could record the answers learners give to create a visualisation of new skills – you could even turn this into a chart! |

| Activity | Instructions | Guidance |
| --- | --- | --- |
| **4** | • Celebrate all that learners have achieved in this chapter.<br>• Wrap up the project by asking: 'How did your new skills help you solve a problem and present a solution?' (Page 86). | • Encourage learners to think about how having visual data not only helped them present their solution, but also gave them confidence as they were backed by data.<br>• Learners should also reflect on their data collection skills and how they improved each time throughout the chapter. |

**Build on this:**

Learners could share insights with the wider community, perhaps at a community meeting or by getting community members to come into the school. If the school has a newsletter, learners could write an article to share their findings.

# Chapter 1: Our digital world

## Chapter 1.1: Messages through time

### Task A: The correct order

| | |
|---|---|
| Messengers on horseback | 3 |
| Email | 5 |
| Cave paintings | 1 |
| Letters through the post | 4 |
| Text messages | 6 |
| Carrier pigeon | 2 |
| Mobile phone video calling | 7 |

### Task B: Decode the messages

1.

| **Coded message** | **Decoded message** |
|---|---|
| 9<br><br>12 15 22 5<br><br>3 15 13 16 21 20 5 18 19 | I<br><br>LOVE<br><br>COMPUTING |
| 8 5 12 12 15<br><br>23 15 18 12 4 | HELLO<br><br>WORLD |
| 13 9 3 18 15 2 9 20 | MICROBIT |

## Chapter 1.3 Engage safely

### Task B: Nicknames

1. RedPenguins
2. There are a few possibilities here – they should incorporate 'blue' and 'remote-controlled cars', for example: 'Ilovebluecars' or 'BlueCar'.

## Chapter 1.4: Think first

### Task A: Levels of privacy

1.

| | Private | Public |
|---|---|---|
| Sending a text message to a friend | ✓ | |
| Adding a comment to a project someone has made in Scratch | | ✓ |
| Sending a picture to a group chat | ✓ | |
| Adding a comment to someone's video online | | ✓ |

# Chapter 2: Content creation

## Chapter 2.1: Improve your typing skills

### Task A: Find the home keys

The home keys will usually be 'F' and 'J', but this will vary on international layouts.

### Task B: Home row

1. On a UK keyboard: ASDFGHJKL. This will vary on international layouts.

### Task D: Home row words

1. Examples: dad, sad, flask, glad, Sal, ask, shall

## Chapter 2.2: Text and shortcuts

### Task A: Spell check and grammar check

1. I have a balloon red and a bicicle green.
2. We see the sun at night. We see the mooon during the day.
3. An spider has six legs and a insect has eight legs.

The spell and grammar check does not deal with factual accuracy.

## Task B: Keyboard shortcuts

1A.  Ctrl + X

1B.  Ctrl + C

1C.  Ctrl + V

1D.  Ctrl + S

1E.  Ctrl + Z

1F.  Ctrl + Y

## Task C: Fix the errors

1.

a. I have a balloon red and a bicycle green.

b. We see the sun at night. We see the moon during the day.

c. A spider has six legs and an insect has eight legs.

## Task E: Word processors and presentation software

1. False
2. False
3. True

# Chapter 3: Create with code 1

## Chapter 3.1: Word search patterns

### Task A: Complete a word search

| | | | | | | | | | | | | | | | | | |
|---|---|---|---|---|---|---|---|---|---|---|---|---|---|---|---|---|---|
| V | K | G | F | E | H | F | Y | X | I | I | T | U | D | I | T | H | Q |
| C | E | F | D | R | H | O | R | I | Z | O | N | T | A | L | L | Y | P |
| I | A | R | N | I | A | X | J | V | Z | A | F | A | G | R | I | D | Z |
| R | Z | L | T | B | V | B | G | L | Z | U | M | S | S | X | F | D | B |
| C | S | V | E | I | O | E | A | H | I | D | D | E | N | Z | I | Y | F |
| L | N | E | R | T | C | A | A | L | A | D | Q | X | I | W | S | B | H |
| E | I | B | A | M | T | A | P | U | Z | Z | L | E | I | S | O | E | P |
| G | Y | O | H | R | R | E | L | L | B | A | H | C | F | Q | L | R | C |
| Z | T | U | U | C | C | A | R | L | U | C | N | M | D | D | F | H | D |
| N | C | K | H | Y | H | H | H | S | Y | J | U | L | U | R | O | H | E |
| L | G | C | D | I | A | G | O | N | A | L | L | Y | Y | D | Q | W | R |
| M | L | O | Z | T | D | S | F | F | Z | K | Z | Y | D | R | B | N | C |

### Task C: Efficient algorithms

1A. B
1B. A
1C. A
1D. B

## Chapter 3.2: Costume change

### Task B: Keep Manpreet warm

1. D

2A. No
3A. Jumper
3B. Repeat 3

## Chapter 3.3: Project design

### Task B: Sprite reset

1. position
2. direction
3. costume
4. size

# Chapter 4: How computers work

## Chapter 4.1 Hardware and software

### Task A. Hardware and software roles

1. B
2. A
3. C

## Chapter 4.2 Input devices

### Task A. Manual input device roles

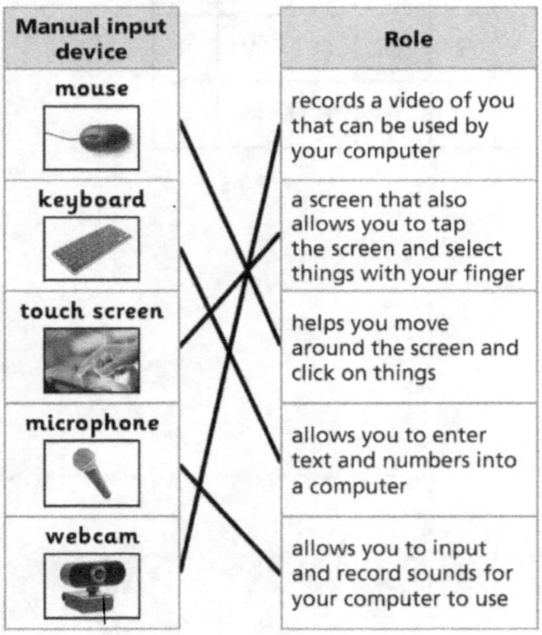


### Task B: Identify the automatic input device

1. barcode reader, light sensor, magnetic stripe reader

# Chapter 5: Create with code 2

## Chapter 5.1: Dance algorithm

### Task A: Is this logical thinking?

1.

| | | Yes | No |
|---|---|---|---|
| A | Guessing answers in a multiple-choice test | | x |
| B | Following the instructions to play a board game | x | |
| C | Planning homework so you complete the next item that you need to hand in | x | |
| D | Eating messy food in your new white top | | x |

## Chapter 5.2: Inputs and outputs with the micro:bit

### Task A: micro:bit inputs and outputs

1.

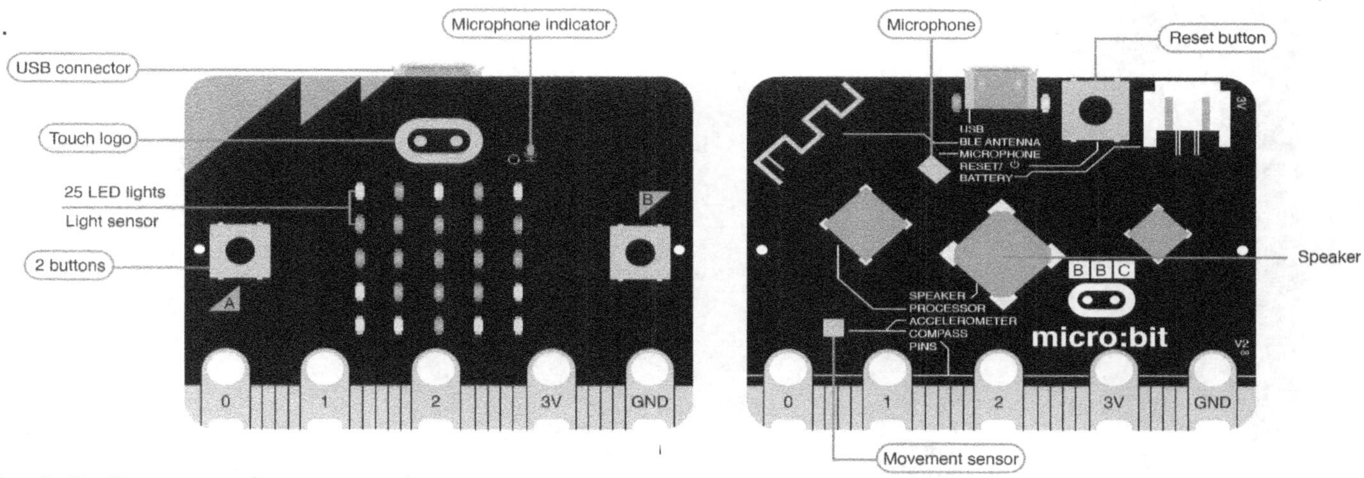


### Task B: Program inputs and outputs

1A. Input component: 'A' button
Output component: Screen
When the 'A' button is pressed the screen will show 1 then 2 then 3 and then nothing (all the LED/lights off).

1B. Input component: 'B' button
Output component: Speaker
When the 'B' button is pressed, the speaker will play musical notes C, D and E with a rest (pause) in between each note.

1C. Input component: Movement sensor
Output component: Screen
When you shake the micro:bit the text "Wow" will scroll across the screen.

## Task C: More program inputs and outputs

1A. Input component: Movement sensor
Output component: Screen
When the micro:bit is tilted to the right the screen will show a heart for one second and then all the LEDs will turn off.

1B. Input component: 'A' and 'B' buttons
Output component: Screen
When the 'A' and 'B' buttons are pressed at the same time, the screen will show an animation of a character jumping four times.

1C. Input component: Light sensor
Output component: Screen
When the light sensor detects that it is dark, the screen will light up.

# Chapter 5.3: Testing for the correct output

## Task A: Match the code to the error

**Code**

Screen should show a square for one second. It should then show a **blank screen for one second** and repeat this four times.

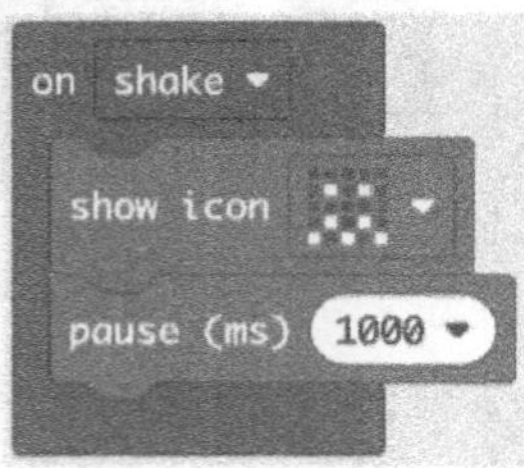


Icon should show for one second and then **disappear**.

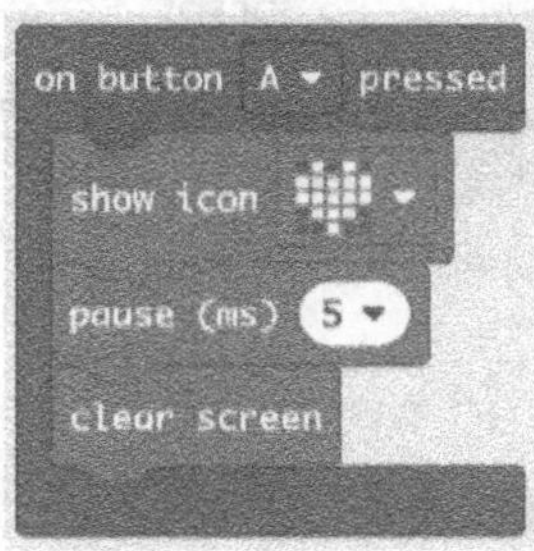


Heart icon should show for five **seconds** and then disappear.

**Error**

missing clear block

seconds instead of milliseconds

missing pause

# Chapter 6: Connect the world

## Chapter 6.1: File types

### Task B: Match the file types to the programs

1.

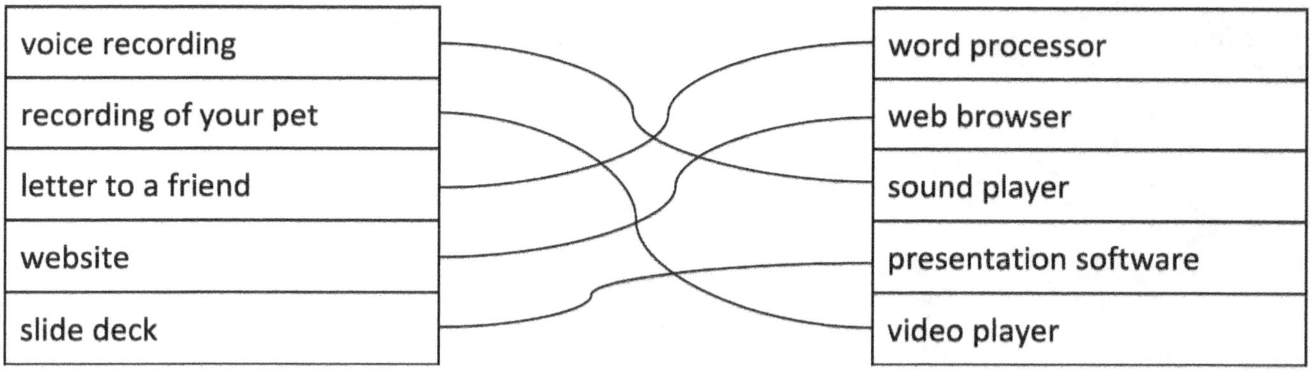


## Chapter 6.2: Common services

### Task B: Share resources

B. The drawing should depict the child using the internet connection from the parent's mobile phone.

C. There are a few ways that this could be drawn. Learners should identify that the photographs should be in a shared location that both the teacher and the headteacher can access.

# Chapter 7: The power of data

## Chapter 7.1: Good data

### Task A: Identifying good data

1A. timely
1B. accurate
1C. complete

## Chapter 7.2: Spreadsheets

### Task A: Answering questions

1A. Rupi
1B. Avi
1C. Simran

## Chapter 7.3: Data charts

### Task A: Chart visualisation

1A. 23.3%
1B. Sunny
1C. Partly cloudy

2A. Rugby Union
2B. Volleyball
2C. 11

## Task B: Reading my chart

1. 4
2. laddu
3. 5

# Websites

## Introduction

| Reference | Website URL |
|---|---|
| IntR1 | https://www.ijcses.org/index.php/ijcses/article/view/15 |
| IntR2 | https://static.teachcomputing.org/GBIC-Evaluation-Report-Relevance.pdf?ref=blog.teachcomputing.org&_ga=2.107553479.476986320.1697481282-1576158333.1697481282&_gac=1.128134910.1697481282.CjwKCAjwvrOpBhBdEiwAR58-3OfKz8kY2yaeSXvgqUBZEl1hUVK8D2AgJRenXD4M3pmNRzU0PWZGARoC42oQAvD_BwE |
| IntR3 | https://stemeducationjournal.springeropen.com/articles/10.1186/s40594-023-00434-7 |
| IntR4 | https://dl.acm.org/doi/10.1145/3287324.3287477 |
| IntR5 | https://www.tandfonline.com/doi/abs/10.1080/08993408.2011.579805 |
| IntR6 | 10.1109/MC.2019.2935058 |
| IntR7 | https://scratch.mit.edu. |
| IntR8 | https://www.google.co.uk/docs/about/ |
| IntR9 | https://support.google.com |
| IntR10 | https://www.google.co.uk/slides/about/ |
| IntR11 | https://support.google.com |
| IntR12 | https://www.google.com/sheets/about |
| IntR13 | https://support.google.com |
| IntR14 | https://microbit.org/get-started/gettingstarted/Introduction |
| IntR15 | https://microbit.org/teach/professionaldevelopment |
| IntR16 | https://www.google.co.uk/forms/about/ |

## Chapter 1

| Reference | Website URL |
|---|---|
| C1L4R1 | https://scratch.mit.edu/community_guidelines |
| C1L5R1 | https://scratch.mit.edu/community_guidelines |

## Chapter 2

| Reference | Website URL |
|---|---|
| C2L1R1 | https://scratch.mit.edu/projects/932862244/fullscreen/ |
| C2L1R2 | https://scratch.mit.edu/projects/932862516/fullscreen/ |
| C2L1R3 | https://scratch.mit.edu/projects/932862244/fullscreen/ |
| C2L1R4 | https://scratch.mit.edu/projects/932862516/fullscreen/ |

## Chapter 3

| Reference | Website URL |
|---|---|
| C3L2R2 | https://scratch.mit.edu/projects/932882750 |

## Chapter 4

| Reference | Website URL |
|---|---|
| C4L2R1 | https://microbit.org/projects/make-it-code-it/nightlight/ |
| C4L2R2 | https://microbit.org/projects/make-it-codeit/nightlight/ |
| C4L6R1 | Certificate of Participation \| The World's Largest Lesson (globalgoals.org) |

## Chapter 5

| Reference | Website URL |
|---|---|

| | |
|---|---|
| C5L2R1 | https://makecode.microbit.org/device/simulator |
| C5L2R2 | https://microbit.org/getstarted/user-guide/transfer-code-tothe-microbit/ |
| C5L2R3 | https://microbit.org/getstarted/user-guide/transfer-code-tothe-microbit/ |
| C5L3R1 | https://makecode.microbit.org/device/simulator |
| C5L4R1 | S3C5 I love space (microbit.org) |
| C5L4R2 | S3C5 Animation (microbit.org) |
| C5L4R3 | S3C5 Music (microbit.org) |
| C5L4R4 | Nightlight \| micro:bit (microbit.org) |
| C5L4R5 | S3C5 I love space (microbit.org) |
| C5L4R6 | https://makecode.microbit.org/S15353-82043-77680-97423 |
| C5L4R7 | https://makecode.microbit.org/S46057-20322-64346-20579 |
| C5L4R8 | https://microbit.org/projects/makeit-code-it/nightlight/ |
| C5L5R1 | https://makecode.microbit.org/device/simulator |